SECOND EDITION

NUMERICAL MODELING of WATER WAVES

SECOND EDITION

NUMERICAL MODELING of WATER WAVES

Charles L. Mader

CRC PRESS

Boca Raton   London   New York   Washington, D.C.

Library of Congress Cataloging-in-Publication Data

Mader, Charles L.
  Numerical modeling of water waves / Charles L. Mader.—2nd ed.
    p. cm.
  Includes bibliographical references and index.
  ISBN 0-8493-2311-8 (alk. paper)
   1. Water waves—Mathematical model. 2. hydrodynamics. I. Title.

QA927.M29 2004
532'.593—dc22   2004047812

This book contains information obtained from authentic and highly regarded sources. Reprinted material is quoted with permission, and sources are indicated. A wide variety of references are listed. Reasonable efforts have been made to publish reliable data and information, but the author and the publisher cannot assume responsibility for the validity of all materials or for the consequences of their use.

Neither this book nor any part may be reproduced or transmitted in any form or by any means, electronic or mechanical, including photocopying, microfilming, and recording, or by any information storage or retrieval system, without prior permission in writing from the publisher.

The consent of CRC Press LLC does not extend to copying for general distribution, for promotion, for creating new works, or for resale. Specific permission must be obtained in writing from CRC Press LLC for such copying.

Direct all inquiries to CRC Press LLC, 2000 N.W. Corporate Blvd., Boca Raton, Florida 33431.

Trademark Notice: Product or corporate names may be trademarks or registered trademarks, and are used only for identification and explanation, without intent to infringe.

Visit the CRC Press Web site at www.crcpress.com

© 2004 by CRC Press LLC

No claim to original U.S. Government works
International Standard Book Number 0-8493-2311-8
Library of Congress Card Number 2004047812
Printed in the United States of America 1 2 3 4 5 6 7 8 9 0
Printed on acid-free paper

ACKNOWLEDGMENTS

The author gratefully acknowledges the developers of the numerical models described in this monograph — M. L. Gittings of SAIC, R. Weaver, G. Gisler, A. Amsden, C. W. Hirt, F. H. Harlow, B. D. Nichols, L. R. Stein, and D. Butler; and the contributions of J. D. Kershner, R. E. Tangora, A. N. Cox, K. H. Olsen, R. Menikoff, E. M. Kober, B. Gonzales, J. C. Solem, J. G. Hills, R. D. Cowan, B. G. Craig, D. Venable, G. E. Seay, and A. L. Bowman of the Los Alamos National Laboratory.

The author wishes to recognize the contributions of Dr. H. Loomis, Dr. W. Adams, Dr. D. Cox, Dr. D. W. Moore, Mr. G. Curtis, Dr. J. Miller, Dr. D. Walker, Dr. S. Furumoto, Dr. L. Spielvogel, Dr. A. Malahoff, Dr. W. Dudley, G. Nabeshima, and S. Lukas of the University of Hawaii.

The author wishes to recognize the collaborations with Dr. E. Bernard of the Pacific Marine Environmental Laboratory, Professor George Carrier of Harvard University, Professor Zygmunt Kowalik of the University of Alaska, Dr. Hermann Fritz of the University of Georgia, Professor Valery Kedrinskii of the Institute of Hydrodynamics, Novosibirsk, Russia, Mr. Tom Sokolowski of the West Coast and Alaska Tsunami Warning Center, Mr. Bruce A. Campbell, Mr. Dennis Nottingham of PN&D, Dr. T. S. Murty of Baird & Associates, Dr. David Crawford of Sandia and Dr. Rob T. Sewell of R. T. Sewell Associates.

This book was reviewed and determined to be unclassified by the Los Alamos National Laboratory group S-7 and given the designation LA-UR-03-8710.

THE AUTHOR

Dr. Charles L. Mader is President of Mader Consulting Co. with offices in Honolulu, Hawaii, Los Alamos, New Mexico and Vail, Colorado. He is an Emeritus Fellow of the Los Alamos National Laboratory. He is the author of *Numerical Modeling of Water Waves*, *Numerical Modeling of Explosives and Propellants*, *Numerical Modeling of Detonations* and four Los Alamos Dynamic Material Properties data volumes.

Dr. Mader is a Fellow of the American Institute of Chemists, Editor of the Science of Tsunami Hazards Journal and is listed in *Who's Who in America* and in *Who's Who in the World*. His web site is http://www.mccohi.com.

INTRODUCTION

This second edition of *Numerical Modeling of Water Waves* describes the technological revolution that has occurred in numerical modeling of water waves during the last decade. A CD-ROM with many of the FORTRAN codes of the numerical methods for solving water wave problems and computer animations of the problems that have been solved using the codes is included with the book. Several PowerPoint presentations describing the modeling results are on the CD-ROM. It will be called the NMWW CD-ROM in the rest of the book.

The objective of this book is to describe the numerical methods for modeling water waves that have been developed primarily at the Los Alamos National Laboratory over the last four decades and to describe some examples of the applications of these methods. Some of the applications of the numerical modeling methods were performed while the author was working at the Joint Institute for Marine and Atmospheric Research at the University of Hawaii, some as Mader Consulting Co. research projects, and the rest as a Fellow of the Los Alamos National Laboratory.

Although the two- and three-dimensional numerical methods for modeling water waves had been available in the 1980's for several decades, they had seldom been used. A major obstacle to their use was the need for access to large and expensive computers. By the 1980's, inexpensive personal computers were adequate for many applications of these numerical methods.

In this book, the basic fluid dynamics associated with water waves are described. The common water wave theories are reviewed in Chapter 1. A computer code called *WAVE* for personal computers that calculates the wave properties for Airy, third-order Stokes, and Laitone solitary gravity waves is available on the NMWW CD-ROM.

The incompressible fluid dynamics model used for shallow water, long waves is described in Chapter 2. A computer code for personal computers using the shallow water model is called *SWAN* and is available on the NMWW CD-ROM. The *SWAN* code is used to model the 1946, 1960 and 1964 earthquake generated Hilo, Hawaii tsunamis, the 1964 Crescent City, California tsunami and the 1994 underwater landslide generated Skagway, Alaska tsunami.

The two-dimensional incompressible Navier-Stokes model used for solving water wave problems is described in Chapter 3. A computer code for personal computers using the two-dimensional Navier-Stokes model is called *ZUNI* and is available on the NMWW CD-ROM. The *ZUNI* code is used to model tsunami wave propagation and flooding. It is also used to model the effect of underwater barriers on tsunami waves.

The three-dimensional incompressible Navier-Stokes model for solving water wave problems of any type is described in Chapter 4. The equations used in the computer code called *SOLA-3D* are described and the FORTRAN code is on the NMWW CD-ROM. The *SOLA-3D* code is used to model the 1975 Hawaiian tsunami and the 1994 Skagway tsunami by water surface cavities generated by underwater landslides.

It is surprising that most academic and government modelers of water waves have chosen to use shallow water or other incompressible models of limited validity for modeling water waves and not use the incompressible Navier-Stokes model since the first edition of this book was published. The severe limitations of the shallow water model are described in Chapter 5.

The generation of water waves by volcanic explosions, conventional or nuclear explosions, projectile and asteroid impacts require the use of the compressible Navier-Stokes model. The numerical models and codes for solving such problems have recently become available as part of the Accelerated Strategic Computer Initiative program. A computer code for solving one, two and three dimensional compressible problems called *SAGE, NOBEL or RAGE* is described in Chapter 6 and some of its remarkable capabilities are presented. These include modeling of the KT Chicxulub asteroid impact and the modeling of the largest historical tsunami which occurred July 8, 1958, at Lituya Bay, Alaska. The modeling of the Lituya Bay impact landslide generated tsunami and the flooding to 520 meters altitude is described.

A color videotape (VTC–86–4) lecture featuring computer generated films of many of the applications discussed in this book is available from the Los Alamos National Laboratory library. Several web sites have been established that contain additional information and computer movies of the problems described in this book. The major sites are http://www.mccohi.org and
http://t14web.lanl.gov/Staff/clm/tsunami.mve/tsunami.html.

CONTENTS

CHAPTER 1 – WATER WAVE THEORY
1A. The Equations of Fluid Dynamics 1
1B. Water Wave Descriptions 9
1C. Airy Waves 16
1D. Laitone Solitary Waves 19
1E. Stokes Waves 22
1F. The Wave Code 29

CHAPTER 2 – THE SHALLOW WATER MODEL
2A. The Shallow Water Equations 31
2B. The Finite-Difference Equations 37
2C. Applications Without Flooding 42
2D. Modeling Hilo, Hawaii Tsunamis 45
2E. Modeling Flooding Around Buildings 61
2F. The 1964 Crescent City Tsunami 70
2G. The 1994 Skagway Tsunami 77

CHAPTER 3 – THE TWO-DIMENSIONAL NAVIER-STOKES MODEL
3A. The Two-Dimensional Navier-Stokes Equations 97
3B. The Finite-Difference Equations 99
3C. Application to Tsunami Wave Propagation 108
3D. Application to Underwater Barriers 120
3E. Waves from Cavities 125

CHAPTER 4 – THE THREE-DIMENSIONAL NAVIER-STOKES MODEL
4A. The Finite-Difference Equations 135
4B. Application to Tsunami Wave Formation 153
4C. The 1994 Skagway Tsunami 163

CHAPTER 5 – EVALUATION OF INCOMPRESSIBLE MODELS FOR MODELING WATER WAVES
5A. Tsunami Wave Generation 165
5B. Tsunami Wave Propagation 176
5C. Tsunami Wave Flooding 183

CHAPTER 6 – MODELING WAVES USING
  COMPRESSIBLE MODELS
6A. The Three-Dimensional Compressible Model 200
6B. The Lituya Bay Mega-Tsunami 203
6C. Water Cavity Generation 215
6D. Asteroid Generated Tsunamis 233
6E. KT Chicxulub Event 245

CD-ROM CONTENTS 255

AUTHOR INDEX 265

SUBJECT INDEX 269

1

WATER WAVE THEORY

1A. The Equations of Fluid Dynamics

The equations of fluid dynamics are representations of the laws of conservation of mass, momentum, and energy as applied to a fluid. The general equations are often called the Navier-Stokes equations. In addition, we need an equation of state to describe the properties of the fluids. The equation of state expresses the fact that the pressure is everywhere a function of the density and the energy per unit of mass. For the water wave theories we will describe in this chapter, the fluid will be considered to be incompressible and the initial density of each element of fluid remains constant. The numerical models of fluid dynamics for compressible fluids are described in the Los Alamos monograph *Numerical Modeling of Detonations*[1], and in Chapter 6.

The equations of fluid dynamics are described by Landau and Lifshitz[2] as a branch of theoretical physics in the classic volume 6 of their *Course of Theoretical Physics*.

The equations of fluid dynamics are treated as a branch of applied mathematics in Batchelor's[3] monograph entitled *An Introduction to Fluid Dynamics*.

A classic description of the equations of fluid dynamics as a branch of oceanography is Blair Kinsman's[4] textbook *Wind Waves*.

Any of these exhaustive treatments of the equations of fluid dynamics are recommended if the reader wishes to reaffirm his faith in the efficacy of Newtonian mechanics and the fluid continuum.

In this monograph, we will first derive the common water wave description starting with the three-dimensional equations of fluid dynamics for viscous, compressible flow.

WATER WAVE THEORY
Chap. 1

The Nomenclature

g_x, g_y, g_z	acceleration due to gravity in x, y, or z direction
I	internal energy
P	pressure
V	specific volume
U_x	particle velocity in x direction
U_y	particle velocity in y direction
U_z	particle velocity in z direction
ρ	density
q	viscosity
q_x, q_y, q_z	viscosity deviators
S_x, S_y, S_z	total deviators
λ, μ	real viscosity coefficients
C	wave velocity
H	maximum wave height
h	wave height
L	wave length
T	wave period
f	frequency
n	wave number
D	depth

Sec. 1A   THE EQUATIONS OF FLUID DYNAMICS

The three-dimensional partial differential equations for viscous, compressible flow:

Eulerian Conservation of Mass

$$\left(\frac{\partial}{\partial t} + U \cdot \nabla\right)\rho = -\rho \nabla \cdot U$$

or

$$\frac{\partial \rho}{\partial t} + U_x\left(\frac{\partial \rho}{\partial x}\right) + U_y\left(\frac{\partial \rho}{\partial y}\right) + U_z\left(\frac{\partial \rho}{\partial z}\right)$$
$$= -\rho\left(\frac{\partial U_x}{\partial x} + \frac{\partial U_y}{\partial y} + \frac{\partial U_z}{\partial z}\right), \qquad (1.1)$$

because

$$\nabla = \frac{\partial}{\partial x}\widehat{i} + \frac{\partial}{\partial y}\widehat{j} + \frac{\partial}{\partial z}\widehat{k},$$

$$U \cdot = U_x\widehat{i} + U_y\widehat{j} + U_z\widehat{k},$$

and

$$\widehat{i} \cdot \widehat{i} = 1, \ \widehat{i} \cdot \widehat{j} = 0, \ \widehat{i} \cdot \widehat{k} = 0, \text{ etc.}$$

WATER WAVE THEORY

Viscosity with No Shearing Forces

$$q_x = 2\mu \frac{\partial U_x}{\partial x} - \frac{2}{3}\mu \left[\frac{1}{V}\left(\frac{\partial V}{\partial t}\right) \right].$$

$$q_y = 2\mu \frac{\partial U_y}{\partial y} - \frac{2}{3}\mu \left[\frac{1}{V}\left(\frac{\partial V}{\partial t}\right) \right].$$

$$q_z = 2\mu \frac{\partial U_z}{\partial z} - \frac{2}{3}\mu \left[\frac{1}{V}\left(\frac{\partial V}{\partial t}\right) \right].$$

$$q = \left(\lambda + \frac{2}{3}\mu\right)\left(\frac{1}{V}\frac{\partial V}{\partial t}\right).$$

$$S_x = q_x - q.$$

$$S_y = q_y - q.$$

$$S_z = q_z - q.$$

Eulerian Conservation of Momentum

$$\rho\left(\frac{\partial}{\partial t} + U\cdot\nabla\right)U = -\nabla\cdot\sigma + \rho g \ .$$

$$\sigma = \delta_{ij}P - S_{ij} \ .$$

$$\rho\left[\frac{\partial U_x}{\partial t} + U_x\left(\frac{\partial U_x}{\partial x}\right) + U_y\left(\frac{\partial U_x}{\partial y}\right) + U_z\left(\frac{\partial U_x}{\partial z}\right)\right]$$
$$= -\frac{\partial(P - S_x)}{\partial x} + \rho g_x \ . \tag{1.2}$$

$$\rho\left[\frac{\partial U_y}{\partial t} + U_x\left(\frac{\partial U_y}{\partial x}\right) + U_y\left(\frac{\partial U_y}{\partial y}\right) + U_z\left(\frac{\partial U_y}{\partial z}\right)\right]$$
$$= -\frac{\partial(P - S_y)}{\partial y} + \rho g_y \ .$$

$$\rho\left[\frac{\partial U_z}{\partial t} + U_x\left(\frac{\partial U_z}{\partial x}\right) + U_y\left(\frac{\partial U_z}{\partial y}\right) + U_z\left(\frac{\partial U_z}{\partial z}\right)\right]$$
$$= -\frac{\partial(P - S_z)}{\partial z} + \rho g_z \ .$$

Eulerian Conservation of Energy

$$\rho\left(\frac{\partial}{\partial t} + U \cdot \nabla\right) I = -\sigma : \nabla U + \lambda \nabla^2 T ,$$

where

$$\sigma : \nabla U = \sigma_{ji} \frac{\partial U_i}{\partial X_j} \quad \text{and}$$

$$\rho\left[\frac{\partial I}{\partial t} + U_x\left(\frac{\partial I}{\partial x}\right) + U_y\left(\frac{\partial I}{\partial y}\right) + U_z\left(\frac{\partial I}{\partial z}\right)\right]$$

$$= -\left[(P - S_x)\frac{\partial U_x}{\partial x} + (P - S_y)\frac{\partial U_y}{\partial y}\right.$$

$$\left. + (P - S_z)\frac{\partial U_z}{\partial z}\right]. \tag{1.3}$$

If the flow is incompressible so $\rho = \rho_o$, equation 1.1 becomes

$$\frac{\partial U_x}{\partial x} + \frac{\partial U_y}{\partial y} + \frac{\partial U_z}{\partial z} = 0 ; \tag{1.4}$$

equation 1.2 becomes, without viscosity,

$$\frac{\partial U_x}{\partial t} + U_x\left(\frac{\partial U_x}{\partial x}\right) + U_y\left(\frac{\partial U_x}{\partial y}\right) + U_z\left(\frac{\partial U_x}{\partial z}\right)$$

$$= -\frac{1}{\rho_o}\frac{\partial P}{\partial x} + g_x , \tag{1.5a}$$

Sec. 1A THE EQUATIONS OF FLUID DYNAMICS

$$\frac{\partial U_y}{\partial t} + U_x\left(\frac{\partial U_y}{\partial x}\right) + U_y\left(\frac{\partial U_y}{\partial y}\right) + U_z\left(\frac{\partial U_y}{\partial z}\right)$$
$$= -\frac{1}{\rho_o}\frac{\partial P}{\partial y} + g_y, \text{ and} \qquad (1.5b)$$

$$\frac{\partial U_z}{\partial t} + U_x\left(\frac{\partial U_z}{\partial x}\right) + U_y\left(\frac{\partial U_z}{\partial y}\right) + U_z\left(\frac{\partial U_z}{\partial z}\right)$$
$$= -\frac{1}{\rho_o}\frac{\partial P}{\partial z} + g_z; \qquad (1.5c)$$

and equation 1.3 becomes, without viscosity,

$$\frac{\partial I}{\partial t} + U_x\left(\frac{\partial I}{\partial x}\right) + U_y\left(\frac{\partial I}{\partial y}\right) + U_z\left(\frac{\partial I}{\partial z}\right)$$
$$= -\frac{1}{\rho_o}\left[P\frac{\partial U_x}{\partial x} + P\frac{\partial U_y}{\partial y} + P\frac{\partial U_z}{\partial z}\right]. \qquad (1.6)$$

Viscosity terms may be included in equation 1.5 as follows. If shearing forces are included and the fluid is incompressible, then

no shear   with shear

$$q_x = 2\mu\frac{\partial U_x}{\partial x} = \mu\left(\frac{\partial U_x}{\partial x} + \frac{\partial U_x}{\partial y} + \frac{\partial U_x}{\partial z}\right),$$

$$q_y = 2\mu\frac{\partial U_y}{\partial y} = \mu\left(\frac{\partial U_y}{\partial x} + \frac{\partial U_y}{\partial y} + \frac{\partial U_y}{\partial z}\right),$$

no shear   with shear

$$q_z = 2\mu \frac{\partial U_z}{\partial z} = \mu \left(\frac{\partial U_z}{\partial x} + \frac{\partial U_z}{\partial y} + \frac{\partial U_z}{\partial z} \right), \text{ and}$$

$$q = 0.0,$$

so equations 1.5 become

$$\frac{\partial U_x}{\partial t} + U_x \left(\frac{\partial U_x}{\partial x} \right) + U_y \left(\frac{\partial U_x}{\partial y} \right) + U_z \left(\frac{\partial U_x}{\partial z} \right)$$
$$= -\frac{1}{\rho_o} \frac{\partial P}{\partial x} + g_x + \frac{\mu}{\rho_o} \left(\frac{\partial^2 U_x}{\partial x^2} + \frac{\partial^2 U_x}{\partial y^2} + \frac{\partial^2 U_x}{\partial z^2} \right), \quad (1.7a)$$

$$\frac{\partial U_y}{\partial t} + U_x \left(\frac{\partial U_y}{\partial x} \right) + U_y \left(\frac{\partial U_y}{\partial y} \right) + U_z \left(\frac{\partial U_y}{\partial z} \right)$$
$$= -\frac{1}{\rho_o} \frac{\partial P}{\partial y} + g_y + \frac{\mu}{\rho_o} \left(\frac{\partial^2 U_y}{\partial x^2} + \frac{\partial^2 U_y}{\partial y^2} + \frac{\partial^2 U_y}{\partial z^2} \right), \quad (1.7b)$$

and

$$\frac{\partial U_z}{\partial t} + U_x \left(\frac{\partial U_z}{\partial x} \right) + U_y \left(\frac{\partial U_z}{\partial y} \right) + U_z \left(\frac{\partial U_z}{\partial z} \right)$$
$$= -\frac{1}{\rho_o} \frac{\partial P}{\partial z} + g_z + \frac{\mu}{\rho_o} \left(\frac{\partial^2 U_z}{\partial x^2} + \frac{\partial^2 U_z}{\partial y^2} + \frac{\partial^2 U_z}{\partial z^2} \right). \quad (1.7c)$$

1B. Water Wave Descriptions

A useful description of water waves has been given by Bernard Le Mehaute in reference 5. Parts of that report are included in the following description of water waves. The theories for unsteady free surface flow subjected to gravitational forces include motions called water waves. They are also called gravity waves. Other gravity waves include motions in other fluids such as atmospheric motions.

A great variety of water waves exists. Water wave motions include storm waves generated by wind in the oceans, flood waves in rivers, seiche or long-period oscillations in harbor basins, tidal bores or moving hydraulic jumps in estuaries, waves generated by a moving ship in a channel, tsunami waves generated by earthquakes, and waves generated by explosions near or under the water.

Mathematically, a general solution does not exist for gravity waves and approximations must be made for even simple waves. One of the important problems in water wave theory is to establish the limits of validity of the various solutions that are due to the simplifying assumptions. The mathematical treatments of the water wave motions use all the mathematical physics dealing with linear and nonlinear problems. The main difficulty in the study of water wave motion is that the free surface boundary is unknown.

Water wave motions are so varied and complex that any attempt at classification may be misleading. Any definition corresponds to idealized situations that never occur. For example, a purely two-dimensional motion never exists. It is a convenient mathematical concept that is physically best approached in a tank with parallel walls. Boundary layer effects and transverse components still exist although they are small and may be neglected in many applications of the theory.

Essentially, two kinds of water waves exist, oscillatory waves and translatory waves. In an oscillatory wave, the transportation of fluid or mass transportation does not occur. The wave motion is then analogous to the transverse oscillation of a rope. A translatory wave involves a transport of fluid in the direction in which the wave travels. For example, a moving hydraulic jump such as a tidal bore is a translatory wave.

An oscillatory wave can be progressive or standing. Consider a disturbance $H(x,t)$ such as a free surface elevation traveling along the OX axis at a velocity C. The characteristics of a progressive wave remain identical for an observer traveling at the same speed and in the same direction as the wave (Figure 1.1). Where h can be expressed as a function of $(x - Ct)$ instead of (x, t), a "steady-state" profile is obtained. $h(x - Ct)$ is the general expression for a progressive wave of steady-state profile traveling in the positive OX direction at a constant velocity C. Where the progressive wave is moving in the opposite direction, its mathematical form is expressed as a function of $(x + Ct)$. The definition of wave velocity C for a non-steady-state profile is unphysical, because each "wave element" travels at its own speed, thereby causing wave deformation.

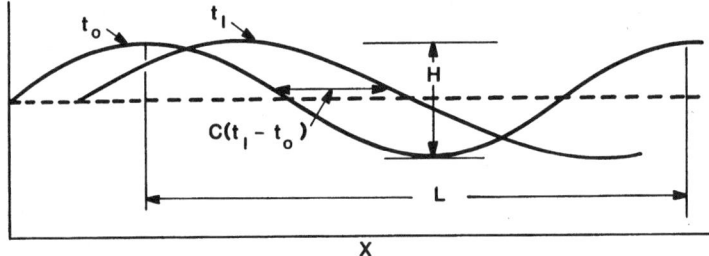

Fig. 1.1. A Progressive Wave.

The simplest case of a progressive wave is the wave defined by a sine or cosine curve such as

$$h = \frac{H}{2} \sin n(x - Ct)$$

or

$$h = \frac{H}{2} \cos n(x - Ct).$$

Such a wave is called a harmonic wave, where $H/2$ is the amplitude and H the wave height.

Sec. 1B WATER WAVE DESCRIPTIONS

The distance between the wave crests is the wave length L, and $L = CT$, where T is the wave period. The wave number $n = 2\pi/L$ is the number of wave lengths per cycle. The frequency is $f = 2\pi/T$. Hence, the previous equations can be written

$$h = \frac{H}{2} \sin 2\pi \left(\frac{x}{L} - \frac{t}{T} \right)$$

or

$$h = \frac{H}{2} \cos 2\pi \left(\frac{x}{L} - \frac{t}{T} \right).$$

A standing or stationary wave is characterized by its mathematical description as a product of two independent functions of time and distance, such as

$$F = H \sin \frac{2\pi x}{L} \sin \frac{2\pi t}{T},$$

or more generally,

$$F = F_1(x) \cdot F_2(t).$$

A standing wave can be considered as the superposition of two waves of the same amplitude and period traveling in opposite directions. Where the convective terms are negligible, the standing wave motion is defined by a linear addition of the equations for the two progressive waves.

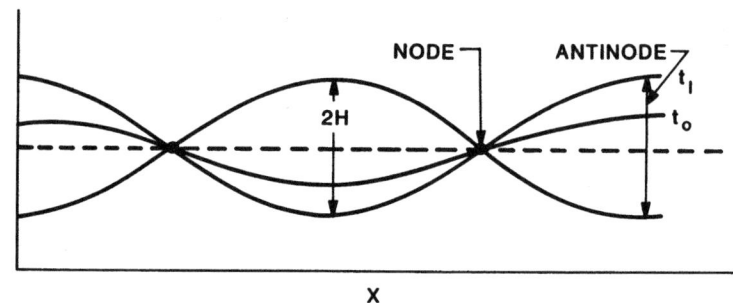

Fig. 1.2. A Standing Wave.

$$\frac{H}{2}\sin\frac{2\pi}{L}(x-Ct) + \frac{H}{2}\sin\frac{2\pi}{L}(x+Ct) = H\sin\frac{2\pi}{L}x\cos\frac{2\pi}{T}t\ .$$

A standing wave generated by an incident wind wave in relatively shallow water ($\frac{D}{L} < 0.05$) is called a seiche. A seiche is a standing oscillation of long period encountered in lakes and harbor basins. The amplitude at the node is zero, and at the antinode, it is H, as shown in Figure 1.2.

Two waves of the same period but different amplitudes traveling in opposite directions can be defined linearly by the summation of $A\sin(x-Ct) + B\sin(x+Ct)$. It can also be considered as the superposition of a progressive wave with a standing wave and is encountered in front of an obstacle that causes a partial reflection of a wave. The amplitude at the node is $N = A + B$ and at the antinode $AN = A - B$. The direct measurement of N and AN yields: $A = \frac{N+AN}{2}$ and $B = \frac{N-AN}{2}$ and the reflection coefficient $R = \frac{N-AN}{N+AN}$.

In a translatory wave, water is transported in the direction of the wave travel. Some examples are

1) A tidal bore or moving hydraulic jump
2) Waves generated by the breaking of a dam
3) A surge on a dry bed
4) An undulated, moving hydraulic jump
5) Solitary waves
6) Flood waves in rivers

In an Eulerian system of coordinates, a surface wave problem generally involves three unknowns: the free surface elevation (or total water depth), the pressure (generally known at the free surface), and the particle velocity.

Since a general method of solution is impossible, a number of simplifying assumptions have been made that apply to a succession of particular cases with varying accuracy.

In general, the method of solution used depends on the relative importance of the convective terms.

However, instead of dealing with these terms directly, it is more convenient to relate them to more accessible parameters. Three characteristic parameters are used.

1) A typical value of the free surface elevation, such as the wave height H
2) A typical horizontal length such as the wave length L
3) The wave depth D

Although the relationships between the convective terms and these three parameters are not simple, their relative values are of considerable help in classifying the water wave theories. As the free surface elevation decreases, the particle velocity also decreases. Thus, when the wave height, H, tends to zero, the convective term, which is related to the square of the particle velocity, is an infinitesimal. Consequently, the convective terms can be neglected and the theory can be linearized.

Thus, the three possible parameters to be considered are

$$\frac{H}{L}, \frac{H}{D}, \text{ and } \frac{L}{D}.$$

The relative importance of the convective terms increases as the values of these three parameters increase.

In deep water (small H/D, and small L/D), the most significant parameter is H/L, which is called the wave steepness. In shallow water, the most significant parameter is H/D, which is called the relative height. In intermediate water depth, a significant parameter that also covers the three cases is $\frac{H}{L}(\frac{L}{D})^3$, and it is called the Ursell parameter.

Depending on the problem under consideration and the range of values of the parameters H/L, H/D, and L/D, four mathematical approaches are used.

1) Linearization
2) Power series
3) Numerical methods
4) Random functions

The simplest cases of water wave theories are the linear wave theories, in which case the convective terms are neglected completely. These theories are valid when H/L, H/D, and L/D are small, i.e., for waves of small amplitude and small wave length in deep water. For the first reason, they are called the "small amplitude wave theory." It is the infinitesimal wave approximation.

The linearization of the basic equation is so suitable to mathematical manipulation that the linear wave theories cover an extreme variety of water wave motions. For example, some phenomena that can be subjected to linear mathematical treatment include the phenomena of wave diffraction and of the waves generated by a moving ship.

The solution can be found as a power series in terms of a small quantity by comparison with the other dimensions. This small quantity is H/L for small L/D because in deep water, the most significant parameter is H/L. It is H/D for large L/D because in shallow water the most significant parameter is H/D.

In the first case (development in terms of H/L), the first term of the power series is given by application of the linear theory. In the second case, the first term of the series is already a solution of nonlinear equations.

The calculation of the successive terms of the series is so cumbersome that these methods are used in a very small number of cases. The most typical case is the progressive periodic wave. In this case, the solution is assumed to be *a priori* that of a steady-state profile, i.e., a function such as $F = f(x - Ct)$, where C is a constant equal to the wave velocity or phase velocity.

The simplification introduced by such an assumption is due to the fact that

Sec. 1B WATER WAVE DESCRIPTIONS

$$\frac{\partial F}{\partial x} = \frac{\partial F}{\partial (x - Ct)}$$

and

$$\frac{\partial F}{\partial t} = -C\frac{\partial F}{\partial (x - Ct)},$$

such that

$$\frac{\partial F}{\partial t} = -C\frac{\partial F}{\partial x}.$$

In such a way, the time derivatives can be eliminated easily and replaced by a space derivative.

Typical examples of such treatments include

1) Power series of H/L or the Stokes waves, valid in deep water. The first term of the series is obtained from linear equations and corresponds to the infinitesimal wave approximation.
2) Power series of H/D: the cnoidal wave or the solitary wave, valid for shallow water. The first terms of the series are obtained as a steady-state solution of already nonlinear equations, but correspond to the shallow water approximation.

However, a steady-state profile may not exist as a solution, in which case the method to be used is often a numerical method of calculation where the differentials are replaced by finite differences. This occurs for large values of H/D and L/D, which correspond to the fact that the nonlinear terms such as $\rho u \frac{\partial u}{\partial x}$ are relatively large by comparison with the linear terms such as $\rho \frac{\partial u}{\partial t}$. This is the case for long waves in very shallow water.

Of course, a numerical method of calculation can be used for solving a linearized system of equations. For example, the relaxation method is used for studying small wave agitation in a basin. Also, an analytical solution of a nonlinear system of equations can be found in some particular cases. Hence, these three methods and the range of application that has been given indicate more of a trend than a general rule.

The three previous methods aim at a fully deterministic solution of the water wave problem. Other descriptions of a sea state generally involve the use of random functions. The mathematical operations (such as harmonic analysis) generally imply that the water waves obey linear laws, which are the necessary requirements for assuming the principle of superposition. Such a method loses its validity for describing the sea state in very shallow water (large values of H/D and L/D).

In hydrodynamics, the water wave theories are generally classified into two families. They are the "small amplitude wave theories" and the "long wave theories."

The small amplitude wave theories are the linearized theories of the first categories of power series, i.e., the power series in terms of H/L.

The long wave theories use the numerical method of solution for the nonlinear long wave equations.

These two families include a number of variations and some intermediate cases with some of the characteristics of both families. For example, the cnoidal wave and the solitary wave are considered as particular cases (steady state) of the long wave theories, because they are nonlinear shallow water waves.

1C. Airy Waves

The linear theory of Airy in Eulerian coordinates gives the essential characteristics of the wave pattern assuming the wave height is infinitesimal. The free surface is sinusoidal as shown in Figure 1.3, particle paths are elliptic and follow a closed orbit (zero mass transport), and lines of equipressure are also sinusoidal. The terms in $(H/L)^2$ are neglected.

Sec. 1C   AIRY WAVES

The long wave or shallow water theory is the same as the theory of Airy, where it is assumed that D/L is small. As a consequence, the formulas are simplified considerably. The pressure is hydrostatic and the horizontal velocity distribution is uniform.

The Airy wave equations used in the *WAVE* code available on the NMWW CD-ROM are given below.

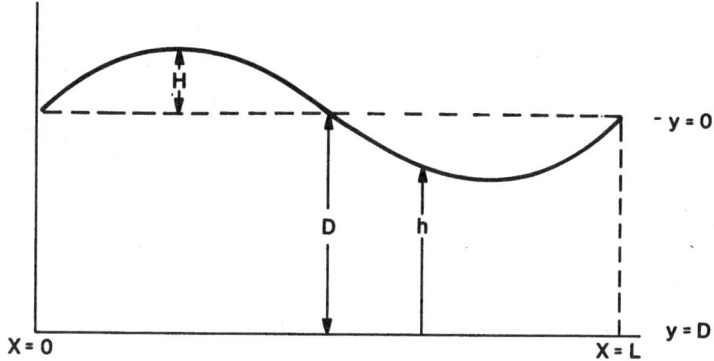

Fig. 1.3. The Airy Wave.

$$C^2 = \frac{gL}{2\pi} \tanh\left(\frac{2\pi D}{L}\right).$$

$$T = \frac{L}{C}.$$

$$P = -\rho g y + \triangle P.$$

$$\triangle P = \rho g H \, \frac{\cosh \frac{2\pi}{L}(y+D) \cos(\frac{2\pi}{L} X)}{\cosh(\frac{2\pi D}{L})}.$$

$$U_x = H \frac{2\pi}{T} \, \frac{\cosh \frac{2\pi}{L}(y+D) \cos(\frac{2\pi}{L} X)}{\sinh(\frac{2\pi D}{L})}.$$

$$U_y = H \frac{2\pi}{T} \frac{\sinh \frac{2\pi}{L}(y+D) \sin(\frac{2\pi}{L}X)}{\sinh(\frac{2\pi D}{L})} .$$

$$h = H \cos\left(\frac{2\pi}{L}X\right) + D ,$$

where h is wave height from the bottom.

$$I = \frac{1}{2}\rho g H^2 .$$

$$G = \frac{C}{2}\left(1 + \frac{4\pi D}{L}/\sinh(4\pi D/L)\right) .$$

Special cases of the Airy wave theory are the deep water ($D/L > 0.5$) and shallow water ($D/L < 0.05$) cases.

For the shallow water case, the Airy waves have straight line "orbits" of length $H\frac{T}{2\pi}\sqrt{\frac{g}{D}}$, and the equations become

$$C = \sqrt{gD} ,$$

$$\triangle P = \rho g H \cos\left(\frac{2\pi}{L}X\right) ,$$

$$U_x = H\sqrt{\frac{g}{D}} \cos\left(\frac{2\pi}{L}X\right) ,$$

Sec. 1D   LAITONE SOLITARY WAVES

$U_y = 0$, and

$G = C$.

For the deep water case, the Airy waves have circular orbits of radius $He^{(2\pi/L)y}$, and the equations become

$$C = \left(\frac{gL}{2\pi}\right)^{1/2} = \frac{gT}{2\pi},$$

$$\triangle P = \rho g H e^{(2\pi/L)y} \cos\left(\frac{2\pi}{L}X\right),$$

$$U_x = H\frac{2\pi}{T} e^{(2\pi/L)y} \cos\left(\frac{2\pi}{L}X\right),$$

$$U_y = H\frac{2\pi}{T} e^{(2\pi/L)y} \sin\left(\frac{2\pi}{L}X\right), \text{ and}$$

$$G = \frac{C}{2}.$$

1D. Laitone Solitary Waves

The cnoidal wave theory of Laitone is mathematically rigorous. At a first order of approximation, the vertical distribution of horizontal velocity is uniform. There is no mass transport. The terms in $(H/D)^2$ are neglected. The wave profile is shown in Figure 1.4.

The theory of Laitone at a second order of approximation gives a nonuniform velocity distribution. There is mass transport. The vertical distribution of mass transport velocity is uniform. The second-order term becomes larger than the first-order term as H/D increases. (H/D is not necessarily a smaller parameter, whereas H/L always is.) The series is nonuniformly convergent. The terms in $(H/D)^3$ are neglected. The results of this theory diverge significantly from experimental results.

The Laitone solitary wave equations used in the WAVE code available on the NMWW CD-ROM are taken from reference 6. The Laitone solitary wave theory is appropriate for H/D less than 0.3. The wave length used for the solitary wave is the length between approximately ± 0.05 maximum wave height, because the wave length is infinite.

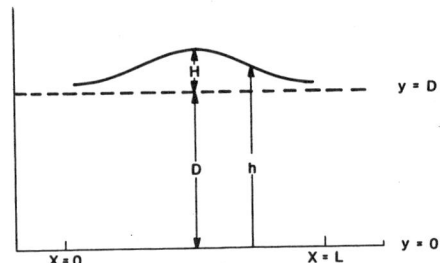

Fig. 1.4. The Laitone Wave.

$$C = \sqrt{gD}\left(1 + \frac{1}{2}\frac{H}{D} - \frac{3}{20}\left(\frac{H}{D}\right)^2\right).$$

The wave length is estimated from

$$L = \frac{2D}{\sqrt{\frac{3}{4}\frac{H}{D}\left(1 - \frac{5}{8}\frac{H}{D}\right)}}.$$

$$T = \frac{L}{C}.$$

Sec. 1D — LAITONE SOLITARY WAVES

$$z = \frac{X}{D}\sqrt{\frac{3\,H}{4\,D}}\left(1 - \frac{5\,H}{8\,D}\right).$$

$$U_x = \frac{\sqrt{gD}\,H}{D}\left[1 + \frac{1}{4}\frac{H}{D} - \frac{3}{2}\frac{H}{D}\frac{y^2}{D^2}\right]\operatorname{sech}^2(z)$$
$$+ \frac{H^2}{D}\left[-1 + \frac{9}{4}\frac{y^2}{D^2}\right]\operatorname{sech}^4(z).$$

$$U_y = \sqrt{gD}\,\sqrt{3}\left(\frac{H}{D}\right)^{3/2}\frac{y}{D}\operatorname{sech}^2(z)\tanh(z)$$
$$\cdot\left[1 - \frac{3}{8}\frac{H}{D} - \frac{1}{2}\frac{H}{D}\frac{y^2}{D^2} + \frac{H}{D}\left(-2 + \frac{3}{2}\frac{y^2}{D^2}\right)\operatorname{sech}^2(z)\right].$$

$$P = \rho g\left\{h - y - \frac{3\,H^2}{4\,D}\left[2\left(\frac{y}{D} - 1\right) + \left(\frac{y}{D} - 1\right)^2\right]\right\}$$
$$\cdot\left[2\operatorname{sech}^2(z) - 3\operatorname{sech}^4(z)\right].$$

$$h = D + H\operatorname{sech}^2(z) - \frac{4}{3}H\left(\frac{H}{D}\right)\operatorname{sech}^2(z)\left(1 - \operatorname{sech}^2(z)\right).$$

1E. Stokes Waves

The theory of Stokes at a second order of approximation is characterized by the sum of two sinusoidal components of period T and $T/2$, respectively. As a result, the wave crest becomes peaked and the troughs become flatter as shown in Figure 1.5. The wave profile can even be characterized by the appearance of a hump in the middle of the wave trough. Similarly, the elliptical particle path is deformed and tends to hump under the crest and flatten under the trough.

In this theory, there is mass transport resulting from irrotationality and nonlinearity. However, phase velocity, wave length, and group velocity are the same as in the linear theories. The terms in $(H/L)^3$ are neglected.

The theory of Stokes at a third order of approximation is characterized by the sum of three sinusoidal terms of period T, $T/2$, and $T/3$, respectively. The same logical results are found. Phase and group velocity exhibit nonlinear corrections. The coefficients of H/L that are functions of D/L tend to infinity when D/L tends to zero, so the theory cannot be used in very shallow water. (The series is nonuniformly convergent.) The terms in $(H/L)^4$ are neglected.

The theory of Stokes at a fifth order of approximation is the sum of five sinusoidal terms. The coefficients of $(H/L)^n$ are functions of D/L and tend to be large values for $n > 3$ even sooner than in the case of the third-order theory, i.e., for larger values of D/L. Consequently, the fifth-order wave theory is less valid than the third-order wave theory for small values of D/L and cannot be used when $D/L < 0.1$. The terms in $(H/L)^6$ are neglected.

The third-order Stokes wave equations used in the *WAVE* code available on the NMWW CD-ROM are taken from reference 6. The Stokes theory is approximate only for waves where D/L is greater than 0.125 and less than 2.

Sec. 1E STOKES WAVES

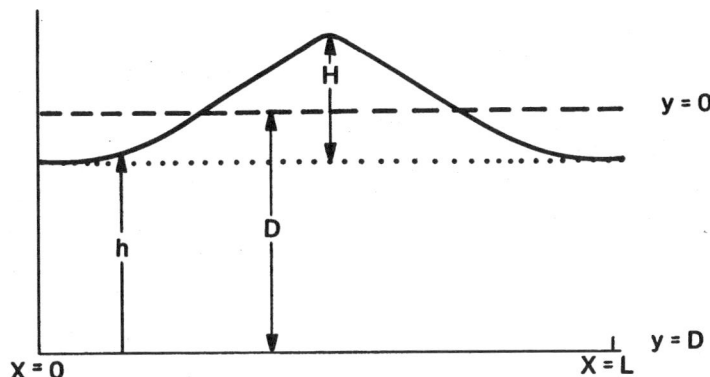

Fig. 1.5. The Stokes Wave.

The wave amplitude (H) is less than half the height and is characteristic of the first component of the wave (a).

$$C^2 = \frac{gL}{2\pi} \tanh \frac{2\pi D}{L} \left[1 + \left(\frac{\pi a}{L}\right)^2 \frac{14 + 4 \cosh^2(\frac{4\pi D}{L})}{16 \sinh^4 \frac{2\pi D}{L}} \right].$$

$$T = \frac{L}{C}.$$

$$h = a \cos \frac{2\pi X}{L} + \frac{\pi a^2}{L} \frac{\left(2 + \cosh\left(\frac{4\pi D}{L}\right)\right) \cosh(\frac{2\pi D}{L})}{2 \sinh^3(\frac{2\pi D}{L})}$$

$$\cdot \cos\left(\frac{4\pi X}{L}\right) + \frac{\pi^2 a^3}{L^2}\left(\frac{3}{16}\right)\left(\frac{1 + 8 \cosh^6(\frac{2\pi D}{L})}{\sinh^6(\frac{2\pi D}{L})}\right)$$

$$\cdot \cos\left(\frac{6\pi X}{L}\right) + D.$$

$$F_1 = \frac{2\pi a}{L} \cdot \frac{1}{\sinh\left(\frac{2\pi D}{L}\right)}$$

$$- \left(\frac{2\pi a}{L}\right)^2 \frac{\left(1 + 5\cosh^2\left(\frac{2\pi D}{L}\right)\right)\cosh^2\left(\frac{2\pi D}{L}\right)}{8\sinh^5\left(\frac{2\pi D}{L}\right)} \cdot$$

$$F_2 = \frac{3}{4}\left(\frac{2\pi a}{L}\right)^2 \cdot \frac{1}{\sinh^4\left(\frac{2\pi D}{L}\right)} \cdot$$

$$F_3 = \frac{3}{64}\left(\frac{2\pi a}{L}\right)^3 \left(\frac{11 - 2\cosh\left(\frac{4\pi D}{L}\right)}{\sinh^7\left(\frac{2\pi D}{L}\right)}\right) \cdot$$

$$U_x = C\left[F_1 \cosh\left(\frac{2\pi(y+D)}{L}\right) \cos\left(\frac{2\pi X}{L}\right)\right.$$

$$+ F_2 \cosh\left(\frac{4\pi(y+D)}{L}\right) \cos\left(\frac{4\pi X}{L}\right)$$

$$\left. + F_3 \cosh\left(\frac{6\pi(y+D)}{L}\right) \cos\left(\frac{6\pi X}{L}\right)\right] \cdot$$

Sec. 1E — STOKES WAVES

$$U_y = C\left[F_1 \sinh\left(\frac{2\pi(y+D)}{L}\right)\sin\left(\frac{2\pi X}{L}\right)\right.$$

$$+ F_2 \sinh\left(\frac{4\pi(y+D)}{L}\right)\sin\left(\frac{4\pi X}{L}\right)$$

$$\left. + F_3 \sinh\left(\frac{6\pi(y+D)}{L}\right)\sin\left(\frac{6\pi X}{L}\right)\right].$$

$$P = -\rho g y + \rho g\left(\frac{H}{2}\right)\left(\frac{\cosh\left(2\pi(\frac{y+D}{L})\right)}{\cosh(\frac{2\pi D}{L})}\right)\cos\left(\frac{2\pi X}{L}\right)$$

$$+ \frac{3}{8}H^2\frac{\pi}{L}\left(\frac{\tanh(\frac{2\pi D}{L})}{\sinh^2(\frac{2\pi D}{L})}\right)\left(\frac{\cosh\left(4\pi(\frac{y+D}{L})\right)}{\sinh^2(\frac{2\pi D}{L})} - \frac{1}{3}\right)$$

$$\cdot \cos\left(\frac{4\pi X}{L}\right) - \frac{H^2}{8}\frac{\pi}{L}\left(\frac{\tanh(\frac{2\pi D}{L})}{\sinh^2(\frac{2\pi D}{L})}\cosh\left(\frac{4\pi}{L}(y+D)\right)\right).$$

$$H = 2a + \frac{2\pi^2}{L^2}a^3\left(\frac{3}{16}\frac{\left(1 + 8\cosh^6(\frac{2\pi D}{L})\right)}{\sinh^6(\frac{2\pi D}{L})}\right).$$

A summary of various wave theory properties taken from reference 5 is shown in Table 1.1. A summary of the properties of waves described in the *WAVE* code is shown in Table 1.2.

WATER WAVE THEORY

TABLE 1.1

Wave Theory Properties

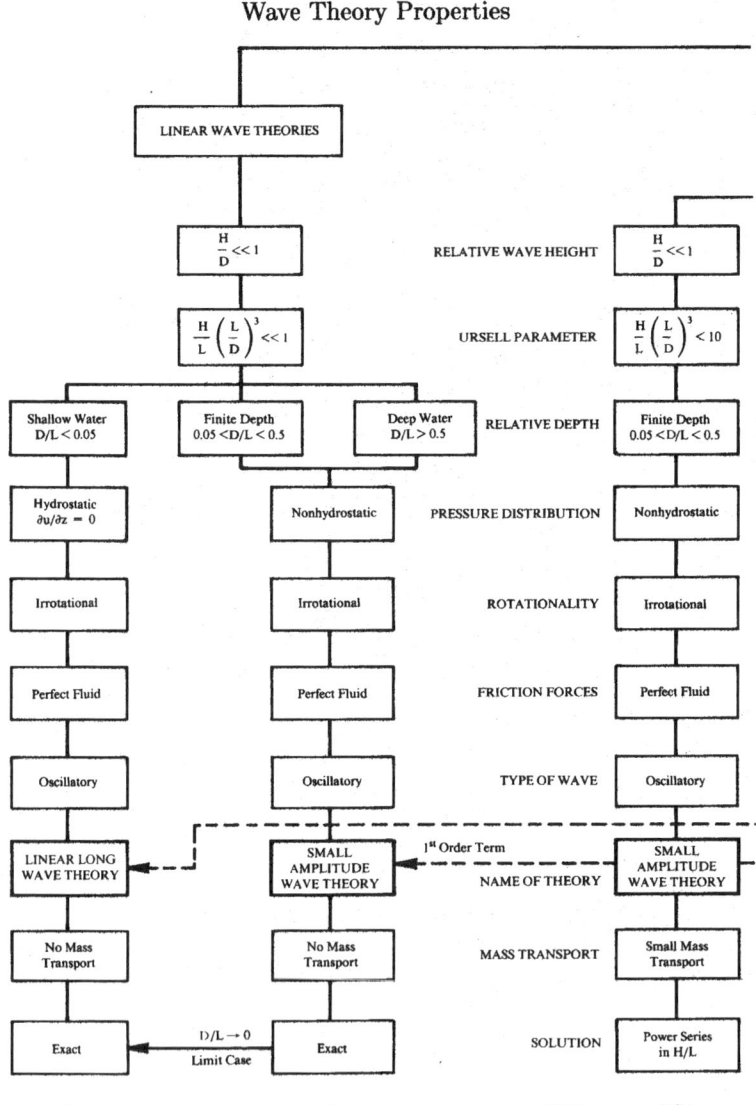

Sec. 1E STOKES WAVES

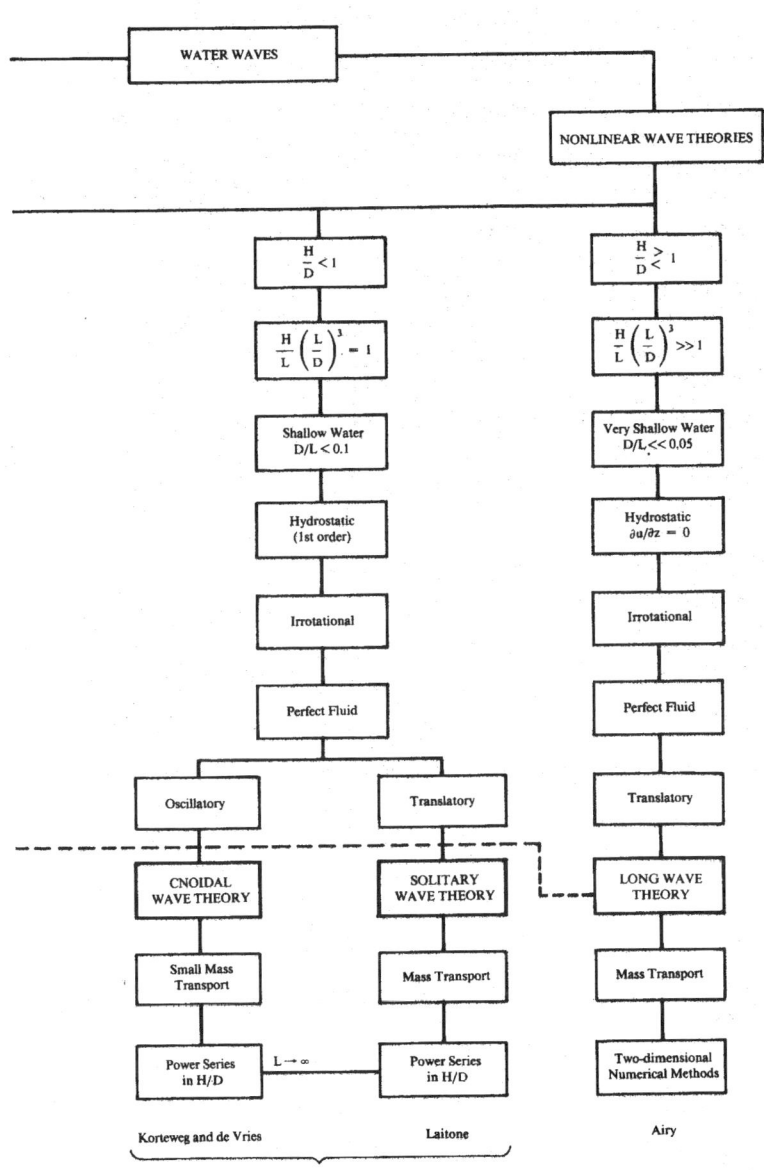

TABLE 1.2

Wave Theory Properties

Properties	Airy	Solitary	Stokes
Incompressible	yes	yes	yes
Linearized	yes	no	no
Nonviscous	yes	yes	yes
Irrotational	yes	yes	yes
Two-Dimensional	yes	yes	yes
Coriolis Effect	no	no	no
Surface Tension	no	no	no
Infinitesimal Wave Height	yes	no	no
Flat Bottom	yes	yes	yes
Sinusoidal Wave	yes	no	no
Exact Solution	yes	no	no
Impermeable Bottom	yes	yes	yes
Wave Form	Sinusoidal	Elongated Trochoidal	Elliptical
Particle Orbit	Closing	Nonclosing	Nonclosing

1F. The Wave Code

The *WAVE* code solves the equations for Airy, third-order Stokes and Laitone solitary gravity waves described in this chapter. It calculates the wave height as a function of wave length and the pressure, and the U velocity and V velocity as a function of wave length and depth. The calculations are performed for the right half of the wave. The *WAVE* code is included on the NMWW CD-ROM in the directory WAVE.

> For Airy Waves: The wave amplitude is the half height, where height is defined as the vertical distance between minimum and maximum of wave surface.
>
> For Solitary Waves: The wave amplitude is the maximum height above still-water level.
>
> For Stokes Waves: The wave amplitude is some value less than the half-height characteristic of the first component of the wave.

For solitary waves, the wave amplitude divided by depth should be less than 0.3. The wave length used for the solitary wave is the length between approximately ± 0.05 maximum wave height since the total wave length is infinite.

For Stokes waves, the depth divided by the wave length should be greater than 0.125 and less than approximately 2.0.

References

1. Charles L. Mader, *Numerical Modeling of Detonations*, University of California Press, Berkeley, California (1979).
2. L. D. Landau and E. M. Lifshitz, *Fluid Mechanics*, Volume 6 of *Course of Theoretical Physics*, Pergamon Press, New York (1959).
3. G. K. Batchelor, *An Introduction to Fluid Dynamics*, Cambridge University Press, London (1967).
4. Blair Kinsman, *Wind Waves: Their Generation and Propagation on the Ocean Surface*, Prentice Hall, Inc., Englewood Cliffs, New Jersey (1965).
5. Bernard Le Mehaute, "An Introduction to Hydrodynamics and Water Waves," ESSA Technical Reports ERL–118–POL 3–1 and 3–2 (1969) and Springer-Verlag, New York (1976).
6. Robert L. Wiegel, *Oceanographical Engineering*, Prentice Hall, Inc., New York (1964).

2

THE SHALLOW WATER MODEL

2A. The Shallow Water Equations

In the classical theory of shallow water, long waves, the vertical acceleration of the fluid particles is neglected because these accelerations are very small with respect to the acceleration of the gravity field. Also, the velocities of the water particles in the z direction may be neglected in dealing with long waves. Thus, all terms containing U_z in equations 1.5 are omitted.

The Cartesian coordinate system is taken in the horizontal plane of the undisturbed water surface. The distance between this reference plane and the bottom is indicated by D, and the distance between this reference plane and the water surface at the considered time is indicated by H.

Following Hansen,[1] vertically averaged velocity components can be introduced according to

$$U_x = \frac{1}{(D+H)} \int_{-D}^{H} u_x dz \, , \text{ and} \tag{2.1}$$

$$U_y = \frac{1}{(D+H)} \int_{-D}^{H} u_y dz \, . \tag{2.2}$$

The distributions of the velocity components U_x and U_y over the vertical at a certain location can be expressed as a function of the averaged velocity by introduction of distribution coefficients,

$$u_x(z) = U_x[1 + u'_x(z)] \, , \text{ and} \tag{2.3}$$

$$u_y(z) = U_y[1 + u'_y(z)] .\qquad(2.4)$$

For these distribution coefficients, the following relations are valid.

$$\int_{-D}^{H} u'_x(z)dz = 0 , \text{ and}\qquad(2.5)$$

$$\int_{-D}^{H} u'_y(z)dz = 0 .\qquad(2.6)$$

The horizontal components of the extraneous forces are the effects of earth rotation and the tide generating force and may be expressed as

$$X = Fu_y + K^{(x)} , \text{ and}\qquad(2.7)$$

$$Y = -Fu_x + K^{(y)} ,\qquad(2.8)$$

where

$F = $ Coriolis parameter, and

$K^{(x)}, K^{(y)} = $ tide generating force.

Sec. 2A THE SHALLOW WATER EQUATIONS

The Coriolis parameter is a function of the latitude and given in Table 2.1 on page 44. The derivation of this parameter is given in Proudman[2] and Dronkers.[3] The Coriolis effect is included as a feature of the $SWAN$ code.

Because the effects of the vertical acceleration and velocity are neglected, Equation (1.5c) may be written

$$\frac{1}{\rho}\frac{\partial p}{\partial z} = Z = -g . \qquad (2.9)$$

In the vertical direction, the external forces Z are the gravity force, the component of the forces induced by earth rotation, and the tide generating force. The latter two are very small compared with gravity and are neglected in this analysis, which concerns shallow water with a depth that is a fraction of the wavelength. It is assumed that the density is uniform. Consequently, in this theory, the pressure is assumed to be hydrostatic and a linear function of the depth,

$$p(z) = \rho g (H - z) + P , \qquad (2.10)$$

where

P = atmospheric pressure, and
ρ = density of water.

The derivatives of the pressure in the horizontal directions now become a function of the water level and the atmospheric pressure,

$$\frac{\partial p}{\partial x} = \rho g \frac{\partial H}{\partial x} + \frac{\partial P}{\partial x} , \text{ and} \qquad (2.11)$$

$$\frac{\partial p}{\partial y} = \rho g \frac{\partial H}{\partial y} + \frac{\partial P}{\partial y}. \qquad (2.12)$$

Integration of equation (1.5a) and equation (1.5b) over the region $z = -D(x,y)$ to $z = H(x,y)$ and introduction of equations 2.1 through 2.8, 2.11, and 2.12 after dividing by $(D+H)$ give

$$\frac{\partial U_x}{\partial t} + U_x \frac{\partial U_x}{\partial x} + U_y \frac{\partial U_x}{\partial y} - FU_y + g \frac{\partial H}{\partial x}$$
$$= -\frac{1}{\rho}\frac{\partial P}{\partial x} + A^{(x)}, \text{ and} \qquad (2.13)$$

$$\frac{\partial U_y}{\partial t} + U_x \frac{\partial U_y}{\partial x} + U_y \frac{\partial U_y}{\partial y} + FU_x + g \frac{\partial H}{\partial y}$$
$$= -\frac{1}{\rho}\frac{\partial P}{\partial y} + A^{(y)}. \qquad (2.14)$$

The terms $A^{(x)}$ and $A^{(y)}$ contain the effects of the tide generating forces and derivatives introduced by the vertical integration. Numerical computation and estimates based upon theoretical considerations show that these terms can generally be omitted in actual computation as shown by Welander[4] and Hansen.[5] It is necessary, however, that the velocity distributions over the vertical be fairly constant.

Up to this point in the discussion, an inviscid fluid has been considered. It can be shown that if viscosity is taken into account, shear-stress terms can be derived from the vertical eddy viscosity in the equation of motion representing effects of wind at the surface and friction at the bottom.

Hansen and Uusitalo also introduce the effects of the horizontal (lateral) eddy viscosity into the equation of motion for computational reasons.[5,6] The terms are very small, and they are neglected in this study.

Sec. 2A THE SHALLOW WATER EQUATIONS

The wind stress is a forcing function in the system of equations, but its dependency upon wind velocity and direction is not discussed here. The bottom stress (τ_b) is proportional to the squared velocity and hence affects the behavior of the long waves considerably. The frictional resistance factor, which is used to establish the relation between the squared velocity and the bottom stress, can be found only by observation. This coefficient depends on the roughness of the bottom, the bottom material, and the depth. Following Dronkers,[3] the relation for flow in one direction is expressed

$$\tau_b = \rho g C^{-2} U |U|,$$

where

ρ = density of the fluid,
U = velocity, and
C = DeChezy coefficient.

As an example, the experimental friction coefficient f, determined from the damping of wind waves, is 0.015 and $f = g/C^2$, so a typical value for C is 25.

Introduction of this bottom-stress term into the two-dimensional system results in

$$\frac{\partial U_x}{\partial t} + U_x \frac{\partial U_x}{\partial x} + U_y \frac{\partial U_x}{\partial y} - FU_y + g\frac{\partial H}{\partial x} + g\frac{U_x(U_x^2 + U_y^2)^{1/2}}{C^2(D+H)}$$
$$= F^{(x)}, \text{ and} \tag{2.15}$$

$$\frac{\partial U_y}{\partial t} + U_x \frac{\partial U_y}{\partial x} + U_y \frac{\partial U_y}{\partial y} + FU_x + g\frac{\partial H}{\partial y} + g\frac{U_y(U_x^2 + U_y^2)^{1/2}}{C^2(D+H)}$$
$$= F^{(y)}, \tag{2.16}$$

where $F^{(x)}$ and $F^{(y)}$ are the forcing functions of wind stress and barometric pressures in the x and y directions, respectively.

In a similar manner, the equation of continuity can be integrated over the vertical. The boundary condition for the free surface is

$$w(H) = \frac{\partial H}{\partial t} + u_x \frac{\partial H}{\partial x} + u_y \frac{\partial H}{\partial y}, \qquad (2.17)$$

and at the bottom,

$$w(-D) + u_x \frac{\partial D}{\partial x} + u_y \frac{\partial D}{\partial y} = 0. \qquad (2.18)$$

With these boundaries, vertical integration of the continuity equation 1.4 results in

$$\frac{\partial H}{\partial t} + \frac{\partial [(D+H)U_x]}{\partial x} + \frac{\partial [(D+H)U_y]}{\partial y} = 0. \qquad (2.19)$$

If we include a bottom motion (R), the terms ($D + H$) in equations 2.15, 2.16, and 2.19 become ($D+H-R$). This derivation is described in more detail in reference 7. Flooding is described using positive values for depths below normal water level and negative values for elevations above normal water level. Only positive values of ($D + H$) terms are allowed. This results in describing both flooding and receding water surfaces.

The long wave theory applies when the depth relative to the wavelength is small and when the vertical component of motion does not influence the pressure distribution that is assumed to be hydrostatic. The long wave theory results in waves that become steeper as they move down a channel, that are too steep as they shoal, and hence, that break too early. This is called the "long wave paradox" and is more serious as the distance and time of interest increases.

Although the "long wave paradox" is a severe limitation on the accuracy of the results, the code can be very useful and inexpensive if applied cautiously to problems appropriate for the method such as tsunami wave formation, propagation, and early shoaling behavior. The limitations of the code for flooding, wave propagation and source modeling are described in Chapter 5.

Many codes have been developed for solving the shallow water, long wave equations. Descriptions and results may be found in references 8–12. The code we will describe is called $SWAN$. The code and its description are available on the NMWW CD-ROM.

2B. The Finite-Difference Equations

The long wave equations (equations 2.15, 2.16, and 2.19) solved by the $SWAN$ code are

$$\frac{\partial U_x}{\partial t} + U_x \frac{\partial U_x}{\partial x} + U_y \frac{\partial U_x}{\partial y} + g \frac{\partial H}{\partial x}$$
$$= F U_y + F^{(x)} - g \frac{U_x (U_x^2 + U_y^2)^{1/2}}{C^2 (D + H - R)},$$

$$\frac{\partial U_y}{\partial t} + U_x \frac{\partial U_y}{\partial x} + U_y \frac{\partial U_y}{\partial y} + g \frac{\partial H}{\partial y}$$
$$= -F U_x + F^{(y)} - g \frac{U_y (U_x^2 + U_y^2)^{1/2}}{C^2 (D + H - R)},$$

and

$$\frac{\partial H}{\partial t} + \frac{\partial (D + H - R) U_x}{\partial x} + \frac{\partial (D + H - R) U_y}{\partial y} - \frac{\partial R}{\partial t} = 0,$$

where

THE SHALLOW WATER MODEL   Chap. 2

U_x = velocity in x direction (i index)
U_y = velocity in y direction (j index)
g = gravitational acceleration (9.8 meter sec^{-2})
t = time
H = wave height above mean water level
R = bottom motion
F = Coriolis parameter
C = coefficient of DeChezy for bottom stress
$F^{(x)}, F^{(y)}$ = forcing functions of wind stress and barometric pressure in x and y direction
D = depth

The wave height H and depth D are taken as cell centered and the velocities are centered at cell boundaries. We change nomenclature where $U = U_x$ and $V = U_y$ to be consistent with the coding convention used in the NMWW CD-ROM. The difference equations at each time step are in order.

$$H_{i,j}^{n+1} = H_{i,j}^n - \Delta t \left[\frac{U_{i+1,j}^n}{\Delta x}(TD1) - \frac{U_{i,j}^n}{\Delta x}(TD2) \right.$$
$$\left. + \frac{V_{i,j+1}^n}{\Delta y}(TV1) - \frac{V_{i,j}^n}{\Delta y}(TV2) \right] + R_{i,j}^{n+1} - R_{i,j}^n \ .$$

$$TD1 = D_{i+1,j} + H_{i+1,j}^n - R_{i+1,j}^n \qquad (U_{i+1,j} < 0) \ .$$

$$TD2 = D_{i,j} + H_{i,j}^n - R_{i,j}^n \qquad (U_{i,j} < 0) \ .$$

$$TV1 = D_{i,j+1} + H_{i,j+1}^n - R_{i,j+1}^n \qquad (V_{i,j+1} < 0) \ .$$

Sec. 2B THE FINITE-DIFFERENCE EQUATIONS

$$TV2 = D_{i,j} + H^n_{i,j} - R^n_{i,j} \qquad (V_{i,j} < 0) .$$

$$TD1 = D_{i,j} + H^n_{i,j} - R^n_{i,j} \qquad (U_{i+1,j} > 0) .$$

$$TD2 = D_{i-1,j} + H^n_{i-1} - R^n_{i-1,j} \qquad (U_{i,j} > 0) .$$

$$TV1 = D_{i,j} + H^n_{i,j} - R^n_{i,j} \qquad (V_{i,j+1} > 0) .$$

$$TV2 = D_{i,j-1} + H^n_{i,j-1} - R_{i,j-1} \qquad (V_{i,j} > 0) .$$

$$U^{n+1}_{i,j} - U^n_{i,j} \quad \triangle t \left[\frac{U_{i,j}}{\triangle x}(TU1) + \frac{TV}{\triangle y}(TU2) \right]$$
$$- g\frac{\triangle t}{\triangle x}[THU] - \triangle t \left[-FV^n_{i,j} - F^{(x)}_{i,j} + S^B_{i,j} \right] .$$

$$TV = 0.25 \left(V_{i,j} + V_{i,j+1} + V_{i-1,j+1} + V_{i-1,j} \right) .$$

$$TU1 = U_{i+1,j} - U_{i,j} \qquad (U_{i,j} < 0) .$$

$$TU2 = U_{i,j+1} - U_{i,j} \quad (TV < 0).$$

$$TU1 = U_{i,j} - U_{i-1,j} \quad (U_{i,j} > 0).$$

$$TU2 = U_{i,j} - U_{i,j-1} \quad (TV > 0).$$

$$THU = H_{i,j} - H_{i-1,j}.$$

$$V_{i,j}^{n+1} = V_{i,j}^n - \Delta t \left[\frac{TU}{\Delta x}(TV1) + \frac{V_{i,j}}{\Delta y}(TV2) \right]$$
$$- g\frac{\Delta t}{\Delta y}[THV] - \Delta t \left[FU_{i,j}^n - F_{i,j}^{(y)} + S_{i,j}^A \right].$$

$$TU = 0.25\left(U_{i,j} + U_{i+1,j} + U_{i,j-1} + U_{i+1,j-1}\right).$$

$$TV1 = V_{i+1,j} - V_{i,j} \quad (TU < 0).$$

$$TV2 = V_{i,j+1} - V_{i,j} \quad (V_{i,j} < 0).$$

$$TV1 = V_{i,j} - V_{i-1,j} \quad (TU > 0).$$

Sec. 2B   THE FINITE-DIFFERENCE EQUATIONS

$$TV2 = V_{i,j} - V_{i,j-1} \qquad (V_{i,j} > 0).$$

$$THV = H_{i,j} - H_{i,j-1}.$$

$$S^B_{i,j} = gU^n_{i,j} \left[{U^n_{i,j}}^2 + {V^n_{i,j}}^2 \right]^{1/2} / C^2(D_{i,j} + H^n_{i,j}).$$

$$S^A_{i,j} = gV^n_{i,j} \left[{U^n_{i,j}}^2 + {V^n_{i,j}}^2 \right]^{1/2} / C^2(D_{i,j} + H^n_{i,j}).$$

The cell whose indices are (1,1) is the southwest corner of the grid, so left, right, bottom, and top refer to the west, east, south, and north sides, respectively. The grid has an artificial boundary so the actual grid spans the indices $(((i,j), i = 2, ip+1), j = 2, jp+1)$.

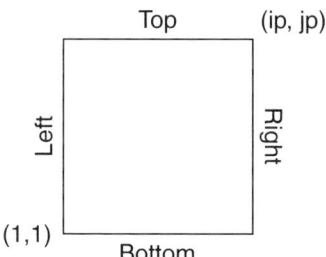

The depth grid should fill the grid plus the artificial boundaries [i.e., a grid dimensioned $(ip+2, jp+2)$].

To obtain stable numerical solutions, the finite-difference equations must not have negative diffusion. The difference equations are stable but require that the time step be kept near the maximum, or the numerical results will become smeared. When choosing the time increment, the time step must also be sufficiently small so a signal takes more than one time step to traverse one grid cell. The Coriolis parameter F may be found from Table 2.1 for any degree of latitude or one can use the Coriolis treatment included in the code.

Boundary conditions of piston, continuum, or reflective are available in the *SWAN* code. The problem mesh is surrounded by a layer of boundary cells whose H, U, and V are prescribed by setting the H of the boundary cell to the nearest neighbor for continuum or reflective boundaries and to the calculated value appropriate for the piston boundary. The appropriate U or V boundary velocity is calculated from the prescription for piston boundaries and is set to zero for reflective boundaries and to the nearest neighbor value for continuum boundaries.

The prescription for the boundary velocity for a shallow water Airy wave is included in the *SWAN* code

$$U_x = A \sin\left((B)(TIME)\right),$$

$$A = H\sqrt{\frac{g}{D}}, \; B = 2\pi/T, \; \text{and}$$

$$H = (U_x \sqrt{gD})/g.$$

2C. Applications Without Flooding

The *SWAN* code has been used to study the interaction of tsunami waves with continental slopes and shelves, as described in reference 13. Comparison with two-dimensional Navier-Stokes calculations of the same problems showed similar results, except for short wavelength tsunamis.

The *SWAN* code was used to model the effects of tides on the Musi-Upang estuaries, South Sumatra, Indonesia, by Safwan Hadi.[14] The computed tide and water discharge were in good agreement with experimental data.

Sec. 2C   APPLICATIONS WITHOUT FLOODING

The *SWAN* code was used to model the large waves that were observed to occur inside Waianae harbor under high surf conditions in reference 15. These waves have broken moorings of boats and sent waves up the boat-loading ramps into the parking lot. The numerical model was able to reproduce actual wave measurements. The *SWAN* code was used to evaluate various proposals for decreasing the amplitude of the waves inside the harbor. From the calculated results, it was determined that a significant decrease of the waves inside the harbor could be achieved by decreasing the harbor entrance depth. Engineering companies used these results to support their recommendations for improving the design of the harbor.

The effect of the shape of a harbor cut through a reef on mitigating waves from the deep ocean was studied using the *SWAN* code in reference 16. It was concluded that a significant amount of the wave energy is dissipated over the reef regardless of the design of the harbor. The reef decreased the wave height by a factor of 3. The wave height at the shore can be further decreased by another factor of 2 by a "V"-shaped or parabolic bottom design.

Other examples of applications of the *SWAN* code are presented in reference 17. They include the wave motion resulting from tsunami waves interacting with a circular and triangular island surrounded by a 1/15 continental slope and from surface deformations in the ocean surface near the island. The effects of a surface deformation in the Sea of Japan similar to that of the May 1983 tsunami was modeled.

In the next chapters, we describe several problems where the shallow water, long wave approximation is not appropriate. They include the wave motion resulting from cavities in the ocean surface, the wave profile of a tsunami wave generated by a sea floor displacement, the wave profile of a tsunami wave generated by an underwater landslide, and the damping action of submerged barriers on tsunami waves.

Many other applications of the linear shallow water, long wave model and the nonlinear model are described in references 8–12. Some of the problems described in these references were inappropriate to be modeled by methods that use shallow water, long wave theory. Often it is difficult to evaluate the error introduced by assuming shallow water, long wave theory without the results of

43

a full Navier-Stokes calculation of the identical problem for comparison. In Chapter 5, we describe the results of such comparisons for wave generation, propagation and flooding.

Many of the most destructive tsunamis of the last century were very long period waves and thus can be modeled realistically using the shallow water model. Tsunami waves generated along the Pacific Ocean rim have to be long period waves if they are to present a significant hazard to Hawaii or other locations located long distances from the source.

TABLE 2.1

The Coriolis Parameter

$$F = 2\omega \sin \phi$$

$$\omega = \frac{2\pi}{24^h} = \frac{2\pi}{60 \cdot 60 \cdot 24} = \frac{2\pi}{86400} = 0.727 \cdot 10^{-4}$$

$$F = \text{Table value} \ x \ 10^{-7}$$

	0	1	2	3	4	5	6	7	8	9
0	000	025	051	076	102	128	153	178	203	228
10	254	279	304	329	353	378	403	426	451	476
20	500	522	546	570	594	616	640	663	685	708
30	730	753	775	796	817	839	859	879	900	919
40	939	959	978	996	1015	1032	1051	1069	1085	1102
50	1118	1136	1151	1167	1182	1196	1210	1224	1239	1252
60	1267	1278	1290	1302	1312	1324	1335	1345	1355	1362
70	1372	1380	1389	1398	1404	1411	1418	1423	1429	1433

2D. Modeling Hilo, Hawaii Tsunamis

Introduction

The flooding of Hilo, Hawaii by the tsunamis of April 1, 1946, May 23, 1960 and March 28, 1964, has been numerically modeled using the non-linear shallow water code *SWAN*, including the Coriolis and friction effects. The Hilo study is described in references 18, 19, 20, and the computer movies are on the NMWW CD-ROM in the directories /TSUNAMI.MVE/1960.MVE, 60THEAT.MVE, 90HILO.MVE and HOTEL.MVE.

The modeling of each tsunami generation and propagation across the Pacific Ocean to the Hawaiian Island chain used a 20 minute grid of ocean depths. This furnished a realistic input direction and profile for the modeling of the tsunami interaction with the Hawaiian Islands on a 5 minute grid. The resulting wave profile and direction arriving outside Hilo Bay was used to model the tsunami wave interaction with the bay, harbor and town on a 100 meters grid. Each element of the grid was described by its height above or below sea level and by a DeChezy friction coefficient determined from the nature of the topography.

The 1946 and 1964 tsunamis were generated by earthquakes in Alaska. The 7.5 magnitude 1946 tsunami flooding of Hilo was much greater than the 8.4 magnitude 1964 tsunami. This was reproduced by the numerical model. The directionality of the tsunami from its source was the primary cause for the smaller earthquake resulting in greater flooding of Hilo. The 1960 tsunami was generated by an earthquake in Chile. The observed largest wave was the third bore-like wave. The numerical model reproduced this behavior. The observed levels of flooding for each event were reproduced by the numerical model, with the largest differences occuring in the Reeds Bay area where the local topography is poorly described by a 100 meters grid. The observed levels of flooding at individual locations were not well described by the model. Since the front and back of a building at a particular location has been observed to have flooding levels varying by a factor of two, a higher resolution grid including the buildings is necessary to describe the flooding at individual locations.

THE SHALLOW WATER MODEL   Chap. 2

Modeling

The flooding of Hilo, Hawaii by the tsunamis of 1946, 1960 and 1964 was modeled using the *SWAN* non-linear shallow water code which includes Coriolis and frictional effects. The 20 and 5 minute topography was obtained from the NOAA ETOPO 5 minute grid of the earth. The 100 meters grid topography and friction coefficients were obtained using available USGS and other topographic maps, photographs and reports.

The extent of flooding for each event is well documented; however the flooding at individual locations was strongly observer dependent. Often the reported flooding at individual locations varied by a factor of two between the different observers and whether the front or the back of a building was used to evaluate the flooding at a location. The available data sources were collected, and a range of flooding observed for each location determined.

A Hilo tsunami date was selected and the following calculations performed:

First — A 20 minute grid calculation of the North Pacific (and when required, for the entire Pacific) was performed to model the tsunami generation and propagation to the region of the Hawaiian Island chain. The 20 minute North Pacific grid was from 120 E to 110 W and 10 N to 65 N and 390 by 165 cells. The wave profile arriving in the region of the chain was used to select a realistic input direction and profile for the second step.

Second — A 5 minute grid calculation of the tsunami wave from the first step interacting with the Hawaiian Island chain was performed. For the 1946 and 1964 tsunami, the 5 minute Hawaiian Island grid was from 163 W to 154 W and 18 N to 24 N and 108 by 72 cells. The wave direction and profile arriving in the region of Hilo Bay was used to select a realistic input direction and profile for the third step.

Third — A 100 meters grid calculation of the tsunami wave from the second step interacting with Hilo Bay and Hilo harbor, and the resulting flooding, was performed using the input wave direction and profile from the second step. The 100 meters Hilo Bay was 100 by 168 cells and the lower left grid corner was located at 155 degrees, 5 minutes, 40 sec and 19 degrees, 42 minutes, 45 sec.

Sec. 2D			MODELING HILO, HAWAII TSUNAMIS

The results of the calculations were compared with the available Hilo flooding levels for each event studied.

Tsunami of April 1, 1946

The tsunami of April 1, 1946 was caused by an earthquake of 7.5 magnitude off the Aleutian islands at 53.5 N, 163 W, 12:29 GMT with a second quake at 12:57 GMT. The source was located about 60 miles SW of Scotch Cap, Unimak Island, where the tsunami destroyed the lighthouse and radio towers located more than 30 meters above sea level. Figure 2.1 shows the lighthouse before and after the tsunami.

The tsunami arrived at Hilo about 7:00 a.m. HST, with a small crest followed by a large recession. The third wave was the largest. Using the first measurable half-wave period, the period was determined from the Honolulu tide gauge to be 15 minutes as described by Green[21]. No instrumental record of the Hilo tsunami was made. The tide at the time of the tsunami was 20 cm above MLLW and falling.

The earthquake source was estimated by Furumoto[22] from the travel times to be 100 kilometers wide and 350 kilometers along the trench. Imaginary wavefronts from observation stations were projected back toward the tsunami source. The presumed source was within the region circumscribed by the interacting wavefronts. A source with these dimensions was chosen which would result in a large negative initial wave at Hawaii and a runup of 30 meters near Scotch Cap lighthouse. The source had a sharp dip of 20 meters along the trench on the deep ocean side, and decreased linearly to 0 meters along the shallow ocean side of the source. For a 20 minute grid, a cross section consisted of cells of $-20.0, -15.0, -10.0, -5.0$ meters initial displacement.

The wave arriving north of the Hawaiian Islands was a 1.0 meter high, 1000 sec period wave with an initial negative pulse. It arrived from the North with the highest energy directed at the islands. This wave was used as the source for the Hawaiian Island calculation.

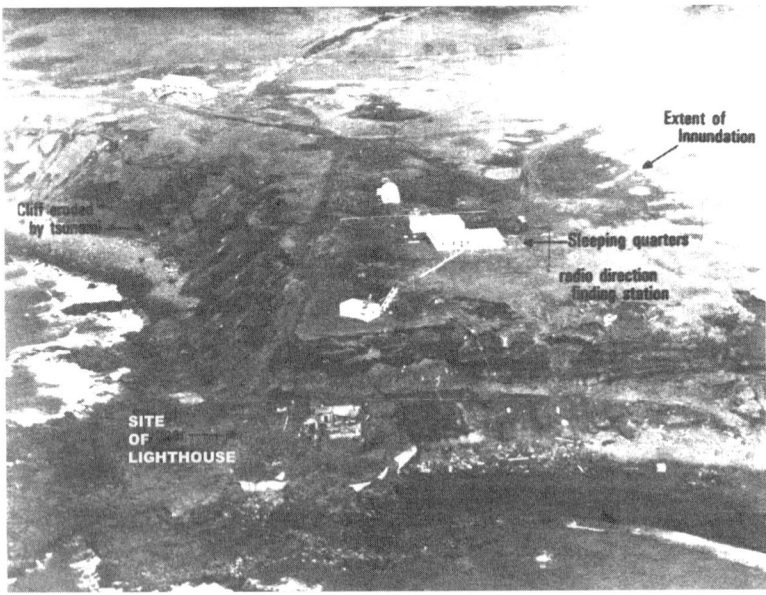

Fig. 2.1. Before and after sequence of tsunami damage at the Scotch Cap lighthouse on Unimak Island, AK. The five-story lighthouse, 50 feet above sea level, was completely destroyed, killing all five occupants. Photograph Credit: U.S. Coast Guard.

The tsunami wave interacted with the Hawaiian Islands and refracted around the island of Hawaii such that the tsunami arrived from the North-East on the Hilo side of the island.

The wave arriving outside Hilo Bay had an initial negative amplitude of 2.0 meters and a 1000 sec period, followed by four 1000 sec period waves with amplitudes of 3 to 4 meters. This wave was used as the source of the Hilo Bay calculation.

The Hilo Bay calculation was performed for the entire bay for the wave starting from the North and from the North-East. The wave from the North refracted into a North-East wave as it interacted with the bay topography. Both calculations gave similar wave interaction and flooding in Hilo harbor.

The flooding was performed using a constant DeChezy friction coefficient of 30, and using the topography determined DeChezy coefficient array shown in Figure 2.2. The roughness coefficients for Hilo harbor and town were determined in the Look Lab Hilo Bay model study in reference 23. The coefficient of 60 is for open smooth areas, 40 for lava like surfaces, 30 for coral and rougher surfaces, 20 for scattered trees and buildings, 10 for buildings and closely spaced trees.

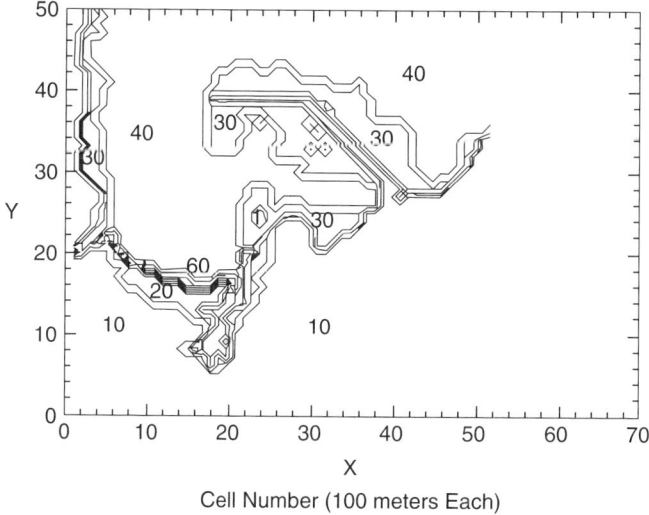

Fig. 2.2. The DeChezy friction coefficients used to describe the topographic roughness in Hilo Bay harbor and town.

THE SHALLOW WATER MODEL   Chap. 2

The calculated and observed inundation limits for Hilo are shown in Figure 2.3. The calculated and observed inundation limits agree to within the 100 meters grid resolution of the numerical model throughout most of the flooded region.

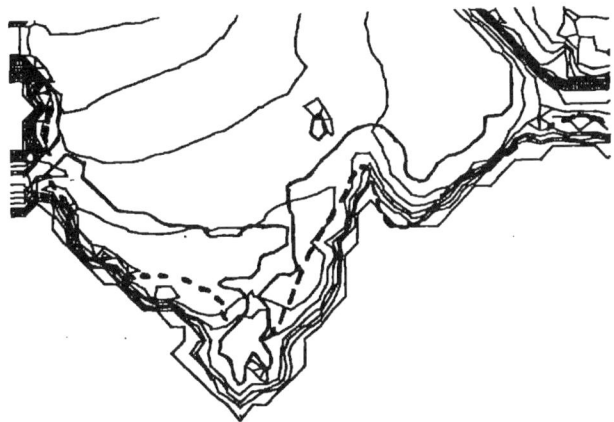

Fig. 2.3. The calculated and observed Hilo inundation limits for the tsunami of April 1, 1946. The observed limit is the heavy dashed line.

The calculated and observed flooding levels at the various harbor locations shown in Figure 2.4 are listed in Table 2.2.

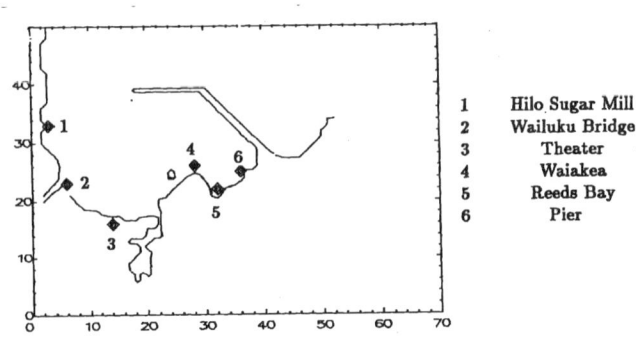

Fig. 2.4. The locations of tsunami inundations in Tables 2.2, 2.3, and 2.4.

TABLE 2.2
April 1, 1946 Tsunami

No	Loc	Obs	Constant Friction	Topo Friction	Look Lab Model
1	Hilo Sugar Mill	7.6	9.5	6.2	
2	Wailuku Bridge	7.3–8.5	8.7	5.8	6.1
3	Theater	6.1–8.5	9.0	6.6	
4	Waiakea	6.7–7.9	6.8	4.8	4.2
5	Reeds Bay	2.4–3.0	8.0	5.7	2.4
6	Pier 2	5.8	7.0	6.0	3.4

The flooding levels are strongly dependent upon the friction. A constant friction model is inadequate to describe either the limits of inundation or the flood levels at individual locations. The shallow water numerical model does not exhibit the observed large variability in flooding at different locations. The 100 meters grid is inadequate to resolve local effects of topography or friction that are important at individual locations. The errors associated with using the shallow water model for flooding are described in Chapter 5. The shallow water model is inadequate to accurately describe the observed flooding at different locations, regardless of the grid size. The hydraulic model reproduces the observed local flooding levels better than the numerical model. The model values were obtained using the hydraulic model of Hilo at Look Laboratories reported in references 23, 24 and 25.

The tsunami was a complete surprise to the residents of Hawaii, resulting in the loss of 159 lives including the man on Pier No. 1 in Hilo Harbor shown in Figure 2.5. The remarkable power of the tsunami is shown. The common definition of a tsunami as a fast rising tide fails to describe the power of the wave and its ability to move large amounts of debris, which become projectiles that can destroy all but the strongest structures.

Fig. 2.5. The 1946 tsunami breaking over Pier No. 1 in Hilo Harbor, Hawaii. The man in the foreground marked by an arrow lost his life. The photograph was taken from the ship *Brigham Victory* which was tied to Pier No. 1. Photograph Credit: NOAA/EDIS.

Tsunami of March 28, 1964

The tsunami of March 28, 1964 was caused by an earthquake of 8.4 magnitude in Alaska near Prince William sound at 61 N, 147.5 W, 03:36 GMT.

The tsunami arrived at Hilo at about 11:15 p.m. HST, with a crest followed by other crests. The second wave was the largest. Using the first measurable half-wave period, the period was determined from the Hilo tide gauge to be 50 minutes. A record of the tsunami off Wake island in 800 feet of water was made by Van Dorn[26]. The wave observed at Wake Island was 15 cm high with a 50 minute period. The tide at the time of the tsunami was at 30 cm above MLLW and rising.

The earthquake source was studied in detail by Plafker[27]. The formation of the tsunami and its interaction (not flooding) with Hilo Bay was modeled by Hwang and Divoky[28,29]. The tsunami was also modeled by Houston, Whalin, Garcia, and Butler[30].

The source was 300 kilometers wide and 800 kilometers long aligned along a SW-NE direction. The source was 7 cells wide in the numerical model. The initial amplitudes from ocean to land had heights of +5.0, +9.0, +10.0, +9.0, +5.0, +1.0, −2.0 meters. This source resulted in a wave at Wake Island similar to that observed by Van Dorn[26].

The wave arriving north of the Hawaiian Islands was much weaker than for the 1946 tsunami. The wave had a profile of a 0.50 meter high half-wave with a period of 4000 sec, followed by a 0.1 meter high half-wave with a period of 2000 sec, then by a 0.25 meter high full wave with a period of 1750 sec.

The tsunami wave interacted with the Hawaiian Islands and refracted around the island of Hawaii such that the tsunami arrived from the North-East on the Hilo side of the island. The wave arriving outside Hilo Bay had an initial positive amplitude of 1.0 meter, 4000 sec period half-wave, followed by a 1.0 meter, 2000 sec period half wave, then by a 1.0 meter, 1750 sec period full wave. This wave was used as the source for the Hilo Bay calculation.

The Hilo Bay calculation was performed for the entire bay for the wave starting from the North and from the North-East. The wave from the North refracted into a North-East wave as it interacted with the bay topography. Both calculations gave similar wave interaction and flooding in Hilo harbor.

The flooding was performed using a constant DeChezy friction coefficient of 30, and using the topography determined DeChezy coefficient array shown in Figure 2.2.

The calculated and observed inundation limits for Hilo are shown in Figure 2.6. They are much smaller than for the April 1, 1946 tsunami. Throughout most of the flooded region the calculated model gave more inundation than was observed.

The calculated and observed flooding levels at various locations in the harbor are listed in Table 2.3. The flooding levels are strongly dependent upon the friction. A constant friction model is inadequate to describe either the limits of inundation or the flood levels at individual locations.

THE SHALLOW WATER MODEL Chap. 2

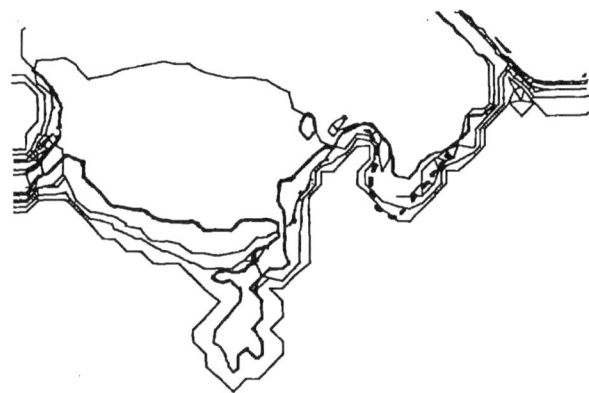

Fig. 2.6. The calculated and observed Hilo inundation limits for the tsunami of March 28, 1964. The observed limit is the heavy dashed line.

TABLE 2.3
March 28, 1964 Tsunami

No	Loc	Obs	Constant Friction	Topo Friction	Look Lab Model
1	Hilo Sugar Mill		2.8	1.9, 2.0	
2	Wailuku Bridge	1.8	3.0	2.0, 2.1	2.7
3	Theater	0.0	2.9	2.0, 2.0	0.0
4	Waiakea	1.5	3.0	2.5, 3.0	1.8
5	Reeds Bay	2.1	3.0	2.6, 3.5	2.3
6	Pier 2	2.4	3.0	2.5, 3.5	2.1

The numerical model does not exhibit the observed large variability in flooding at different locations. The 100 meters grid and the shallow water model are inadequate to resolve local effects of topography or friction that are important at individual locations. The hydraulic model reproduces the observed local flooding levels better than the numerical model.

Sec. 2D   MODELING HILO, HAWAII TSUNAMIS

The 1946 and 1964 tsunamis were generated by earthquakes in Alaska. The 7.5 magnitude 1946 tsunami flooding of Hilo was much greater than the 8.4 magnitude 1964 tsunami. This was reproduced by the numerical model. The directionality of the tsunami from its source was the primary cause for the smaller earthquake resulting in greater flooding of Hilo. The 1946 tsunami wave peak energy was directed toward Hawaii while the 1964 tsunami wave peak energy was directed east of Hawaii toward the Pacific Coast of North America. The large waves observed at Crescent City for the 1964 tsunami and not for the 1946 tsunami are consistent with this directionality difference. In Section 2F, the 1964 tsunami and the flooding of Crescent City will be modeled.

Tsunami of May 23, 1960

This study required development of a full North and South Pacific grid to determine the nature of the wave arriving at Hilo from South America. A one degree grid from 110 E to 65 W and 65 S to 65 N of 185 by 130 cells and a 20 minute grid of 555 by 390 cells was developed. The one degree grid could only resolve the source with a two cell wide source and gave waves with periods two times larger than observed. So the 20 minute grid tsunami wave profiles were used to describe the 1964 tsunami.

The tsunami of May 23, 1960 was caused primarily by an earthquake at 19:11 GMT on May 23, 1960 of 8.5 magnitude occurring near Peru, Chile and centered at 38 N and 73.5 W. The major earthquake was preceeded by two 7.5 magnitude quakes, one at 10:03 and another at 19:10 GMT.

The main tsunami wave crested at Hilo at 12:13 a.m. HST on May 23. The first wave peak was followed by a second peak at 12:46 a.m., then by a third peak (a bore at the harbor entrance) about 20 minutes later which was more than twice as high as the previous waves. This wave was the highest and most destructive tsunami wave in Hilo's history. The remarkable tsunami destructive power is shown in Figure 2.7. The bent parking meters and gutted foundation illustrate the power of the wave. The town of Hilo near the ocean before and after the tsunami is shown in Figure 2.8.

Fig. 2.7. The aftermath of the 1960 tsunami in the Waiakea area of Hilo, Hawaii. Bent parking meters show the direction of the tsunami. Photograph Credit: U.S. Navy.

Fig. 2.8. The town of Hilo facing the ocean before and after the 1960 tsunami. Photograph Credit: U.S. Navy.

After the 1960 tsunami completely destroyed all of Hilo near the ocean, the area was turned into a park. One can still view the foundations of many of the buildings that were destroyed. A visit to the Pacific Tsunami Museum in Hilo, Hawaii is the best way to develop an understanding of what happened to Hilo when the 1960 tsunami occurred. The museum occupies a concrete bank building that survived the tsunami. The museum web site is http://planet-hawaii.com/tsunami.

The tide at 12:07 a.m. HST was at 60 cm above MLLW and increasing. By the time the largest third and fourth waves arrived the tide was cresting at 70 cm above MLLW.

The earthquake was studied by Plafker and Savage[31]. The formation of the tsunami and its propagation across the Pacific Ocean toward Hawaii was numerically modeled by Hwang and Divoky[28,29]. They concluded that peak wave heights occur along a path roughly normal to the major axis of the elongated source region. They suggested that the preferential directivity may account for the severity in Japan of the Chilean tsunami.

The source was 6 cells or 150 kilometers wide and 21 cells or 800 kilometers long aligned along a N-S direction. The source was six cells wide and an initial upward displacement of $1, 2, 4, 6, 4, -2$ meters along the width. The waves arrived at the Hawaiian Island chain from the South-East. The wave had a profile of a 0.225 meter high, 3000 sec half wave followed by a 0.525 meter high, 1500 sec wave and then followed by a 0.49 meter high, 1500 sec wave. The wave that arrived at Johnson Island was similar to the wave reported by Van Dorn[26].

The 5 minute Hawaiian Island grid was from 170 W to 140 W and 15 N to 25 N and 360 by 120 cells. The tsunami wave interacted with the Hawaiian Islands and refracted such that the wave arrived from the east on the Hilo side of the island. The wave arriving outside Hilo Bay had an initial amplitude of 0.4 meter and a period of 3000 sec for 1500 sec, followed by a 1.5 meters, 1500 sec wave, and then by a 2.0 meters, 1500 sec wave.

The flooding was performed using a topography determined DeChezy coefficient. The roughness coefficients for Hilo harbor and town were determined by the Look Lab Hilo Bay model study and are shown in Figure 2.2.

The calculated and observed inundation limits for Hilo are shown in Figure 2.9. The calculated and observed inundation limits agree to within the 100 meters grid resolution of the numerical model throughout most of the flooded region, with the calculated model giving more inundation than observed between Reeds Bay and Pier 2.

The calculated and observed flooding levels at various locations in the harbor are listed in Table 2.4.

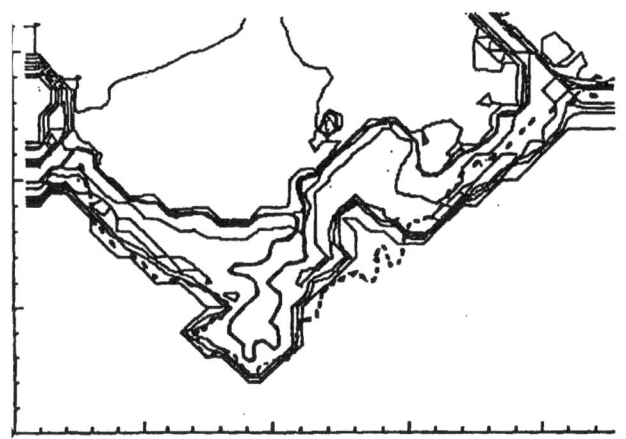

Fig. 2.9. The calculated and observed Hilo inundation limits for the tsunami of May 22, 1960. The observed limit is the heavy dashed line.

TABLE 2.4

May 23, 1960 Tsunami

No	Loc	Obs	Constant Friction	Topo Friction	Look Lab Model
1	Hilo Sugar Mill	4.6–6.1		3.7	6.7
2	Wailuku Bridge	4.3–5.8		4.5	3.8
3	Theater	6.7–8.5		4.5	7.6
4	Waiakea	4.6–6.1		4.2	3.5
5	Reeds Bay	2.7–3.7		4.9	4.1
6	Pier 2	3.6–4.3		5.0	4.4

The numerical model does not exhibit the observed large variability in flooding at different locations. The 100 meters grid and the shallow water model (see Chapter 5) are inadequate to resolve local effects of topography or friction that are important at individual locations. The hydraulic model reproduces the observed local flooding levels better than the numerical model.

Sec. 2D   MODELING HILO, HAWAII TSUNAMIS

The calculated third wave was the largest and steepest in agreement with the observations, although the difference in amplitude between the second and third wave is not as large as observed.

The numerical modeling results support the suggestion of Hwang and Divoky that the preferential directivity accounts for the severity of the Chilean tsunami in Hawaii and Japan. The interaction of the tsunami wave with Hilo Bay resulted in modifying the amplitude of the waves from the second being the largest to the third being the largest, steepest and most bore-like. The wave arriving at Wake Island exhibited none of these characteristics. The interaction of tsunami waves with Hilo Bay is strongly dependent upon their period and their interaction with bay topography and with each other. The relative amplitude and steepness of the waves outside of Hilo Bay may be quite different from the waves flooding the town of Hilo.

The 1960 tsunami modeling movies are on the NMWW CD-ROM in the directory /TSUNAMI.MVE/1960.MVE.

Conclusions

The flooding of Hilo, Hawaii, by the tsunamis of April 1, 1946, May 23, 1960 and March 28, 1964, has been numerically modeled using the non-linear shallow water code *SWAN* including the Coriolis and friction effects. The modeling of each tsunami generation and propagation across the Pacific Ocean to the Hawaiian Island chain followed by modeling of the tsunami interaction with the Hawaiian Islands on a finer grid and then modeling the tsunami wave interaction with the bay, harbor and town using a high resolution grid results in inundation limits that reproduce the essential features of the actual inundation limits.

The 1946 and 1964 tsunamis were generated by earthquakes in Alaska. The 7.5 magnitude 1946 tsunami flooding of Hilo was much greater than the 8.4 magnitude 1964 tsunami. This was reproduced by the numerical model. The directionality of the tsunami from its source was the primary cause for the smaller earthquake resulting in greater flooding of Hilo. The 1960 tsunami was generated by an earthquake in Chile. The observed largest wave was the third bore-like wave. The numerical model reproduced this behavior.

The observed levels of flooding for each event were reproduced by the numerical model with the largest differences occuring in the Reeds Bay area where the slope is shallow and the topography is poorly described by a 100 meters grid. The observed levels of flooding at individual locations were not well described by the model. The shallow water model does not accurately describe flooding for slopes less than 2 percent as described in Chapter 5. Since the front and back of a building at a particular location has been observed to have flooding levels varying by a factor of two, a higher resolution grid including the buildings is needed to describe the flooding at individual locations. The Hilo Theater flooding by the 1960 tsunami is described in Section 2E using a higher resolution 10 meters grid.

Nested Grid Modeling

Kowalik and Whitmore[32] have developed a finite difference technique for solving the shallow water equations which allows for fine numerical grids inside courser grids. Such a method would permit modeling the Hilo tsunamis in a single step. The technique has been used by the West Coast and Alaska Tsunami Warning Center, at Palmer, Alaska, to compute expected tsunami amplitudes off the coastlines of Alaska, the U.S. West Coast, and Hawaii for 15 hypothetical earthquakes in the northwest Pacific ranging from 7.5 to 9.0 moment magnitude by Whitmore and Sokolowski[33].

The development of the nested grid methodology by Kowalik has resulted in the usual duplication of effort by other numerical modelers. The interfaces between the fixed grids of different resolution introduces uncertainities that can result in large errors. What is needed is a methodology of arbitrary continuous mesh refinement with the grids varying only by a factor of two wherever needed similar to that described in Chapter 6 for compressible Navier-Stokes modeling. Considering the large errors associated with the shallow water model discussed in Chapter 5, it is uncertain if further effort to improve the nested grid capability of shallow water codes would be useful.

2E. Modeling Flooding Around Buildings

Hilo Theater

As described in Section 2D, the observed levels of flooding at various locations were not well described using a 100 meters grid. To resolve the flooding at individual locations and around buildings a 10 meters grid of Hilo was developed.

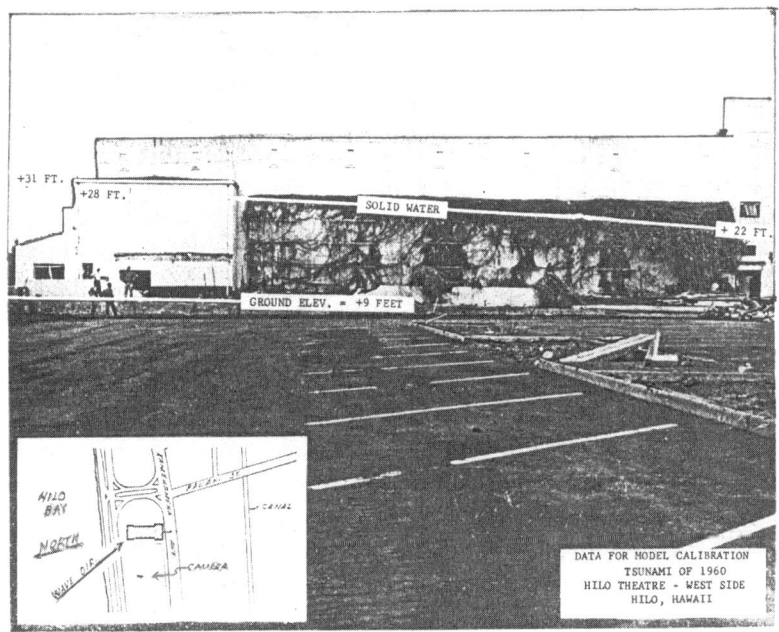

Fig. 2.10. The Hilo Theater after the 1960 tsunami. The solid water line is 28 to 22 feet above sea level. The building is located 9 feet above sea level. The inset shows the location of the theater relative to the shoreline. The photograph was taken from the parking lot on the west side of the building. Note the four men standing on the left side of the photograph.

THE SHALLOW WATER MODEL   Chap. 2

A high resolution 10 meters grid of Hilo was used to model the flooding around Hilo Theater by the 1960 tsunami wave. The results are described in reference 34. The Hilo Theater was located near the shore, in flat and unobstructed terrain 2.7 meters above sea level. The tsunami flooded level reached 8.5 meters at the seaward side and 6.7 meters at the rear of the Hilo Theater. A photograph of the Hilo Theater after the May 1960 tsunami is shown in Figure 2.10.

The grid was 240 by 240 cells of 10 meters on a side. The time step was 0.2 sec. Friction was described as in Section 2D and Figure 2.2. The theater building was described by 3 by 6 cells with a height of 11.9 meters. The third wave was a bore-like wave inside the harbor entrance 3 meters high from the North-West. The initial water level was 1 meter lower than the normal sea level to approximate the second wave recession that occurred before the third wave arrived. The wave had a period of 1500 sec with the first rise occuring in 60 sec to approximate the bore-like wave.

The surface level of the water, land and theater building is shown at various times in Figure 2.11. The contour interval is one meter. The building is 11.9 meters high.

The height of the water at locations near the harbor entrance and near the building as a function of time are shown in Figure 2.12. The front of the theater building (Location 5) was flooded to the 8.5 meters level and the rear of the building (Location 6) to the 8.0 meters level. The rear of the building was calculated to flood to a higher level than observed.

The inundation of the town of Hilo is shown in Figure 2.13. The inundation continues after the maximum flooding of the theater occurs. The maximum level of inundation calculated is similar to that observed from the 1960 tsunami.

The maximum level of flooding observed at Hilo Theater was reproduced by the high resolution numerical model using a realistic description for the 1960 tsunami wave.

Sec. 2E   MODELING FLOODING AROUND BUILDINGS

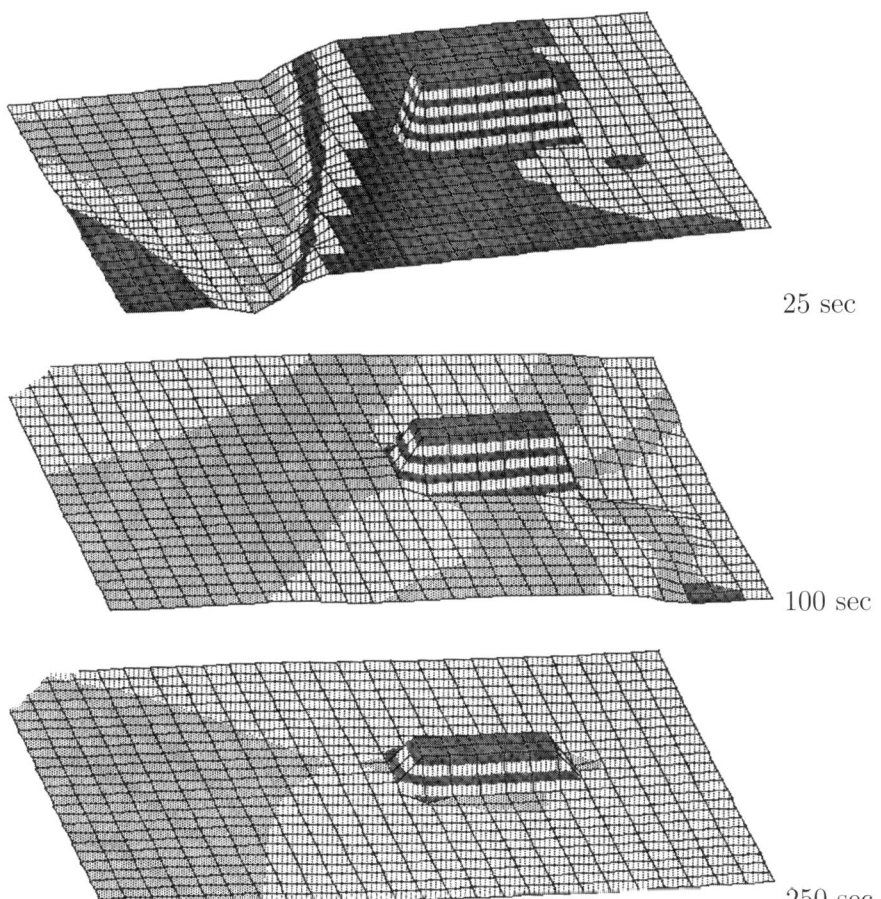

25 sec

100 sec

250 sec

Fig. 2.11. The surface level of the water, land and the Hilo Theater building at 25, 100 and 250 sec. The contour interval is one meter. The building is 11.9 meters above sea level on land that is 2.7 meters above sea level. The initial water surface is 1 meter below normal sea level to approximate the second wave withdrawal before the arrival of the largest third wave.

THE SHALLOW WATER MODEL   Chap. 2

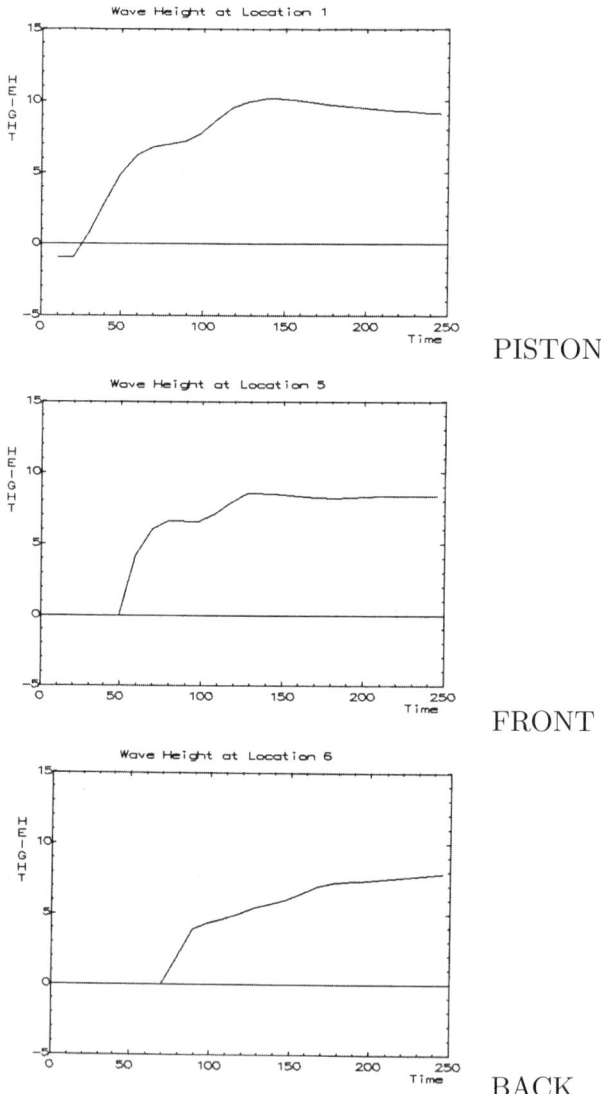

Fig. 2.12. The height of the water at locations near the piston and near the building as a function of time are shown. Location 1 is in the ocean near the piston. Locations 5 and 6 are on the front and the back of the side of the theater shown in Figure 2.10. The initial water surface is 1 meter below normal sea level to approximate the second wave withdrawal before the arrival of the largest third wave.

Sec. 2E   MODELING FLOODING AROUND BUILDINGS

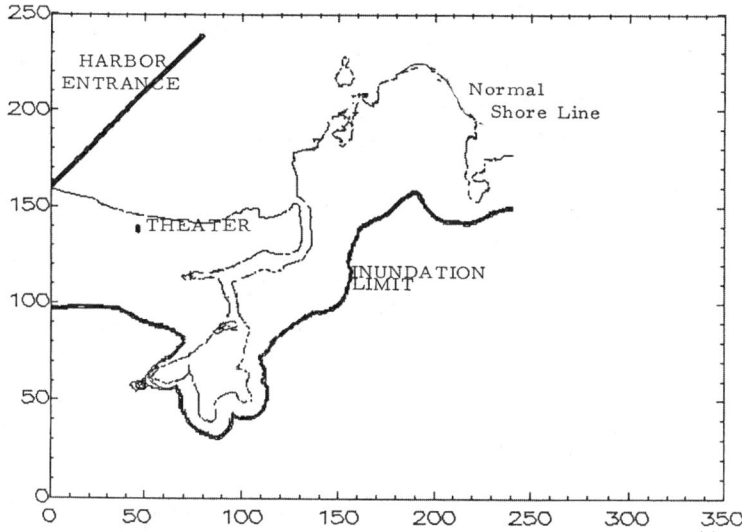

Fig. 2.13. The inundation of the town of Hilo is shown using a 10 meters grid. The inundation continues after the maximum flooding of the theater occurs. The location of the normal shoreline and the theater building is shown. Also shown is the location of the piston boundary at the harbor entrance used to simulate the third 1960 tsunami wave.

In the directory /TSUNAMI.MVE/60THEAT.MVE on the NMWW CD-ROM are computer movies showing the interaction of the 1960 tsunami with the Hilo Theater.

Current Hilo Buildings

Having demonstrated that the observed flooding around the Hilo Theater can be modeled, we can now model a 1960 tsunami and how it would interact with the buildings and topography that exists today.

After the 1960 tsunami, a region of downtown Hilo was filled with earth and used for construction of government buildings and a shopping center. The hope was that the landfill would raise the land and buildings sufficiently above sea level that a 1960 type of tsunami wave would not flood the new construction.

THE SHALLOW WATER MODEL   Chap. 2

The current buildings and hotels as they exist today were added to the 10 meters grid and modeling was performed using a similar model for the 1960 tsunami as used for the Hilo Theater.

The 10 meters resolution Hilo topography with the landfill and buildings as they existed in the 1990's is shown in Figure 2.14. The contour interval is 1 meter in the first frame. A 4 meters high, 1500 sec period wave just outside the harbor simulated the 1960 tsunami wave. Friction was described as in Section 2D and Figure 2.2. The calculated inundation from the tsunami wave is shown in the bottom frame. The region with the large buildings on the right side of the lagoon is the region of the landfill. The landfill appears to be sufficient to prevent flooding of the region from a 1960 type of tsunami.

The computer movies for the tsunami flooding of Hilo with buildings is in the directory /TSUNAMI.MVE/90HILO.MVE on the NMWW CD-ROM.

After the 1960 tsunami, a large development of hotels occurred along the sea shore across from Coconut island and next to Reeds Bay. The hotels were constructed with the first three floors open, with the hope that a 1960 type of tsunami would pass thru the building. The two bottom floors are an automobile garage and the third floor is an open air hotel lobby.

The 10 meters resolution topography with the hotels as they existed in the 1990's is shown in Figure 2.15. From right to left is Coconut Island, the Hilo Hawaiian Hotel, the Hilo Bay Hotel, the Hawaii Naniloa Hotel and Reeds Bay.

The flooding of the hotel region and Banyan Drive by a 4 meters high, 1500 sec tsunami is shown in Figure 2.16. The tsunami is similar to the 1960 tsunami, which was the largest Hilo historical tsunami. The power plant is shown at the top right of the figure, which serves as a timing mark for Hilo tsunamis as the electric clocks stop when a tsunami floods the plant.

The computer movies for the tsunami flooding of the Hilo hotels is in the directory /TSUNAMI.MVE/HOTEL.MVE.

Sec. 2E   MODELING FLOODING AROUND BUILDINGS

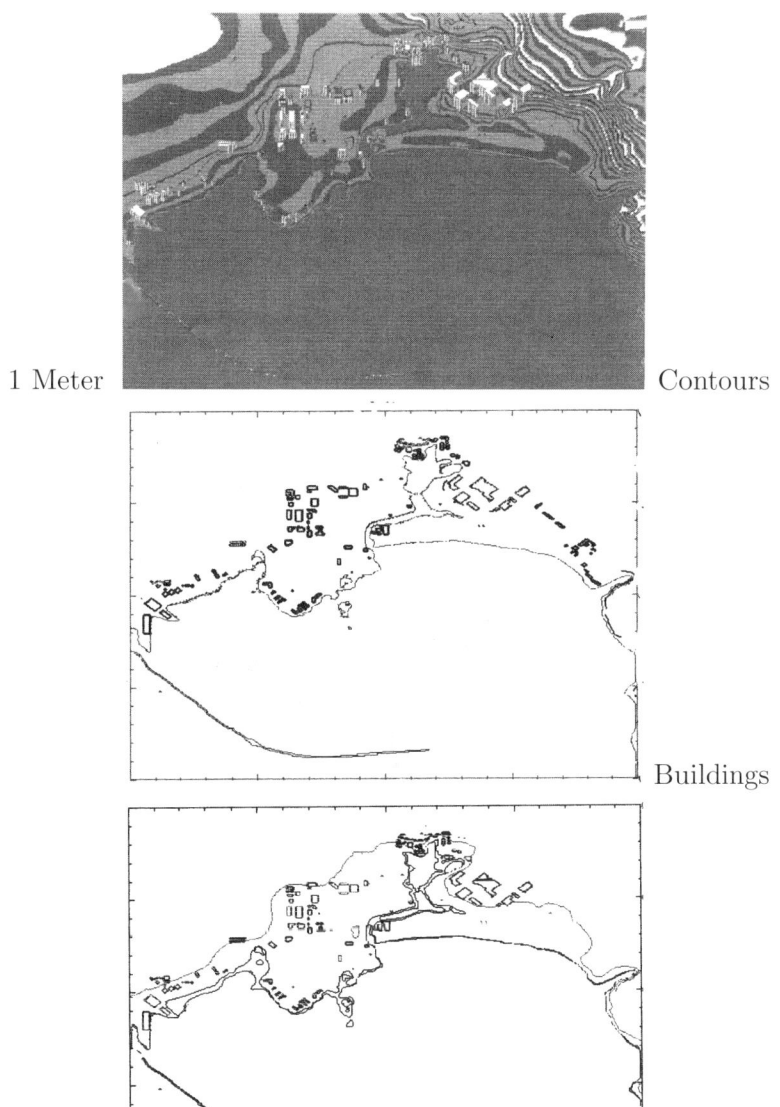

Fig. 2.14. The town of Hilo with buildings after the landfill and construction of government buildings on the landfill. The calculated maximum inundation from a 1960 tsunami wave is shown. The landfill appears to be sufficient to protect from the maximum historical tsunami.

THE SHALLOW WATER MODEL   Chap. 2

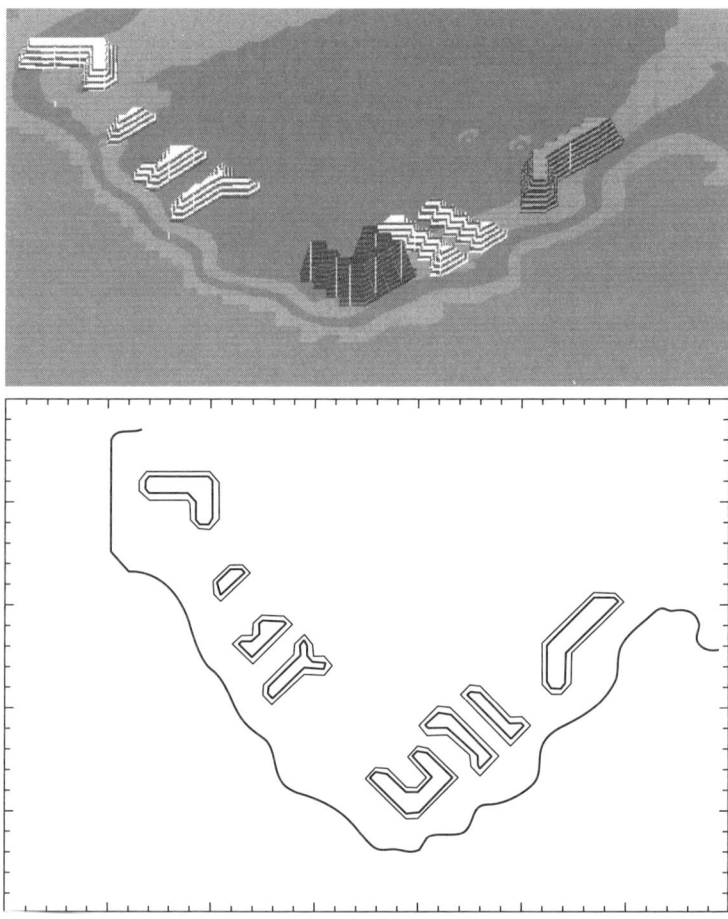

Fig. 2.15. The numerical model for the Banyan Drive Hilo Hotels, with the Hilo Hawaiian Hotel, the Hilo Bay Hotel, the Hawaii Naniloa Hotel from right to left between Coconut Island and Reeds Bay. The contour interval is 1 meter.

Sec. 2E   MODELING FLOODING AROUND BUILDINGS

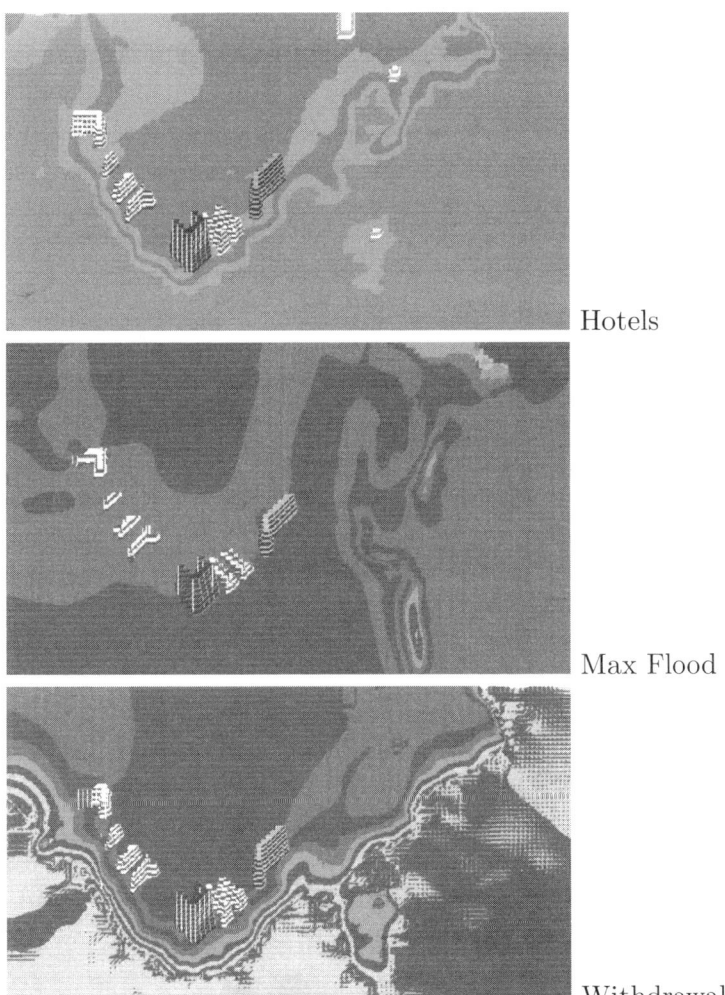

Hotels

Max Flood

Withdrawal

Fig. 2.16. The flooding of the Banyan Drive Hilo Hotels. The times are 0.0, 380 sec and 600 sec for a 4 meter, 1500 sec period wave. Such a wave simulates the maximum historical Hilo tsunami wave.

THE SHALLOW WATER MODEL   Chap. 2

2F. The 1964 Crescent City Tsunami

Introduction

The March 28, 1964 Crescent City tsunami was caused by an Alaskan earthquake. The disaster at Crescent City exceeded all the combined effects in historical time from tsunamis on the U.S. West Coast. In Crescent City there were ten fatalities due to drowning. The port facilities and 29 city blocks containing 172 businesses, twelve house trailers, and 91 homes were damaged or destroyed. Twenty one boats were sunk. The inundation map for Crescent City is shown in Figure 2.17.

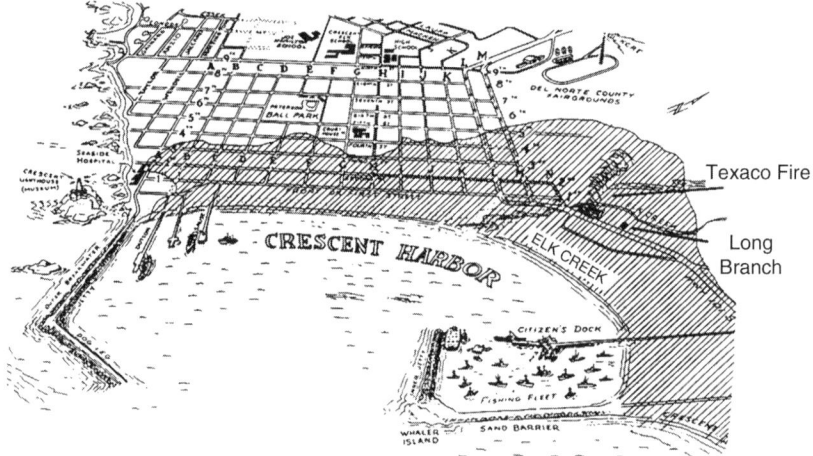

Fig. 2.17. The inundation map for Crescent City for the March 28, 1964 tsunami. Photo Credit: NGDC/NOAA.

The tsunami generation and propagation across the Pacific was modeled using a 20 minute grid for the North Pacific. The wave arriving in the region of the U.S. West Coast was modeled using a 5 minute grid. The wave arriving outside of Crescent City harbor was then modeled using a 25 meters grid of the harbor and town. The Crescent City tsunami study is described in reference 35.

Sec. 2F   THE 1964 CRESCENT CITY TSUNAMI

The model gave approximately the observed maximum area of flooding of Crescent City. The large amount of flooding of Hilo, Hawaii from the 7.6 magnitude 1946 Alaskan earthquake and small amount of flooding from the 8.4 magnitude 1964 Alaskan earthquake at Hilo while extensively flooding Crescent City was reproduced by the numerical model.

Modeling

The flooding of Crescent City, California by the tsunami of 1964 was modeled using the *SWAN* non-linear shallow water code which includes Coriolis and frictional effects described in this chapter. The effect of the tide was included and found to be small.

The 20 and 5 minute topography was obtained from the NOAA ETOPO 5 minute grid of the earth. The 25 meters grid topography was obtained using available USGS and other topographic maps, photographs and reports. The extent of flooding for the event is well documented[29] and shown in Figure 2.17.

The following calculations performed were similar to the study of tsunami wave inundation of Hilo, Hawaii described in Section 2D.

First — A 20 minute grid calculation of the North Pacific was performed to model the tsunami generation and propagation to the region of the West Coast of the U.S. The 20 minute North Pacific grid was from 120 E to 110 W and 10 N to 65 N and 390 by 165 cells. The wave profile arriving in the region of the West Coast was used to select a realistic input direction and profile for the second step.

Second — A 5 minute grid calculation of the tsunami wave from the first step interacting with the West Coast was performed. The 5 minute West Coast grid was from 10 N to 60 N and 240 by 480 cells. The wave direction and profile arriving in the region of Crescent City harbor was used to select a realistic input direction and profile for the third step.

Third — A 25 meters grid calculation of the tsunami wave from the second step interacting with Crescent City harbor and town, and the resulting flooding was performed using the input wave direction and profile from the second step. The 25 square meters Crescent City grid was 160 by 240 cells.

The results of the calculations were compared with the available Crescent City flooding levels for the 1964 tsunami.

The tsunami of March 28, 1964 was caused by an earthquake of 8.4 magnitude in Alaska near Prince William sound at 61 N, 147.5 W, 13:36 GMT.

The modeling of the tsunami wave formed by the earthquake and its interaction with Hilo, Hawaii was described in Section 3D. The earthquake source was studied in detail by Plafker[27]. The source was 300 kilometers wide and 800 kilometers long aligned along a SW-NE direction. The source in the numerical model was 7 cells wide. The initial amplitudes from ocean to land had heights of $+5.0$, $+9.0$, $+10.0$, $+9.0$, $+5.0$, $+1.0$, -2.0 meters. This source resulted in a wave at Wake Island similar to that observed by Van Dorn[24]. The wave observed at Wake Island was 15 cm high with a 50 minute period.

The calculated and observed Hilo inundation limits were much smaller than for the April 1, 1946 tsunami. Both the 1946 and 1964 tsunamis were generated by earthquakes in Alaska. The 7.5 magnitude 1946 tsunami flooding of Hilo was much greater than the 8.4 magnitude 1964 tsunami. This was reproduced by the numerical model. The directionality of the tsunami from its source was the primary cause for the smaller earthquake resulting in greater flooding of Hilo. The 1946 tsunami wave peak energy was directed toward Hawaii while the 1964 tsunami wave peak energy was directed east of Hawaii toward the Pacific Coast of North America as shown in Figure 2.18.

At the northern end of the West Coast of the U.S. the tsunami wave shown in Figure 2.19 has a period of 1500 sec and a half-wave amplitude of about 2.0 meters shown as Location 7 in Figure 2.19. In the deep ocean the wave is coming from the North-West. It refracts as it travels down the coast as shown in Figure 2.20 and is coming from the West as it interacts with the region of Crescent City with heights of 4 meters as shown in Figure 2.21, Locations 6 and 7.

Figure 2.22 shows the calculated and observed inundation limit for Crescent City. The observed inundation limit was described in reference 29. The tsunami wave outside of the harbor had a height of 4.0 meters and a period of 1500 sec. A constant DeChezy friction coefficient of 30 was used.

Sec. 2F   THE 1964 CRESCENT CITY TSUNAMI

The 1964 tsunami arrived just after high tide which contributed to the level of flooding. The tide was included in the model and the ebbing tide decreased the maximum water levels by less than 10 percent.

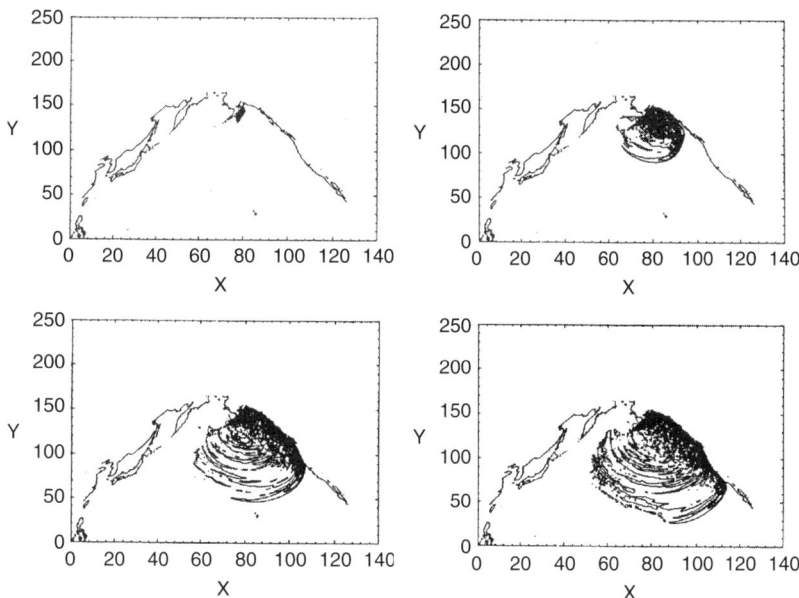

Fig. 2.18. The March 28, 1964 tsunami wave interacting across the North Pacific Ocean. The contour interval is 0.35 meters. The times are 60, 7200, 14000 and 18000 sec and the grid size was 20 minute.

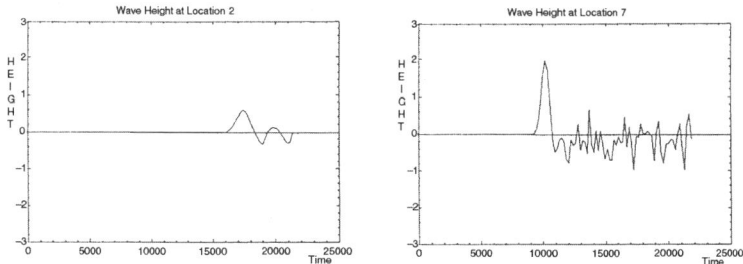

Fig. 2.19. The March 28, 1964 tsunami wave moving across the North Pacific Ocean. Location 2 is east of the island of Hawaii in 301 meters of water. Location 7 is west of Seattle in 2803 meters of water.

73

THE SHALLOW WATER MODEL   Chap. 2

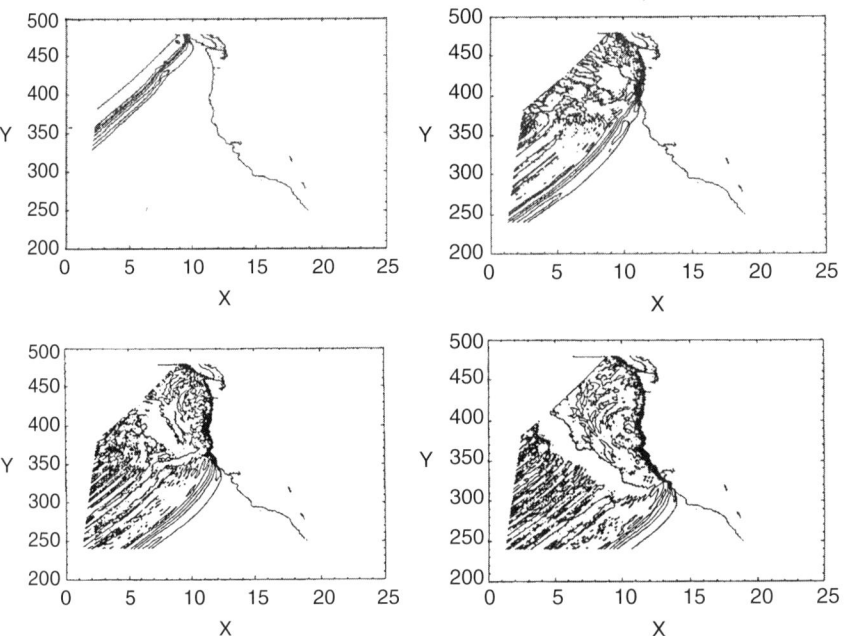

Fig. 2.20. The March 28, 1964 tsunami wave interacting with the U.S. West Coast. The contour interval is 0.70 meter. The times are 15, 4500, 6000 and 7500 sec and the grid size was 5 minutes.

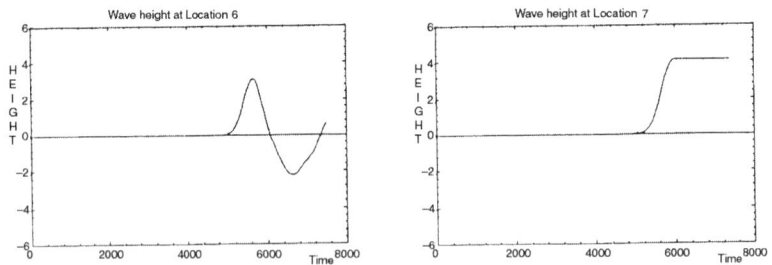

Fig. 2.21. The March 28, 1964 tsunami wave interacting with the U.S. West Coast. Location 6 is outside of Crescent City in 45 meters of water. Location 7 is at the shoreline of Crescent City.

The calculated flooding of Crescent City by the 1964 tsunami is shown in Figure 2.23.

Sec. 2F   THE 1964 CRESCENT CITY TSUNAMI

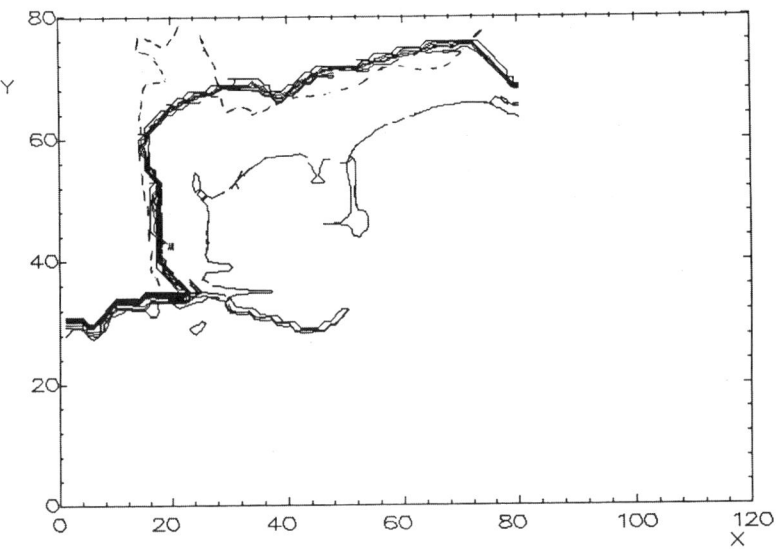

Fig. 2.22. The calculated and observed inundation limits for Crescent City for the March 28, 1964 tsunami. The observed inundation limit is shown as a dashed line.

Conclusions

The flooding of Crescent City, California by the tsunami of March 28, 1964 has been numerically modeled using the non-linear shallow water code $SWAN$ including the Coriolis and friction effects. The modeling of the tsunami generation and propagation across the Pacific Ocean to the U.S. West Coast followed by modeling of the tsunami interaction with the West Coast on a finer grid and then modeling the tsunami wave interaction with Crescent City harbor and town using a high resolution grid results in inundation limits that reproduce the observed inundation limits.

The 1946 and 1964 tsunamis were generated by earthquakes in Alaska. The 7.5 magnitude 1946 tsunami flooding of Hilo was much greater than the 8.4 magnitude 1964 tsunami while the reverse was observed for Crescent City, California. This was reproduced by the numerical model. The directionality of the tsunami from its source was the primary cause for the smaller 1946 earthquake resulting in greater flooding of Hilo than the large 1964 earthquake.

75

THE SHALLOW WATER MODEL   Chap. 2

The directionality of the tsunami was also the primary cause for the extensive flooding of Crescent City by the 1964 event.

In the directory /TSUNAMI.MVE/1964.MVE on the NMWW CD-ROM are computer movies showing the propagation of the 1964 Alaskan tsunami to Crescent City and the resulting inundation.

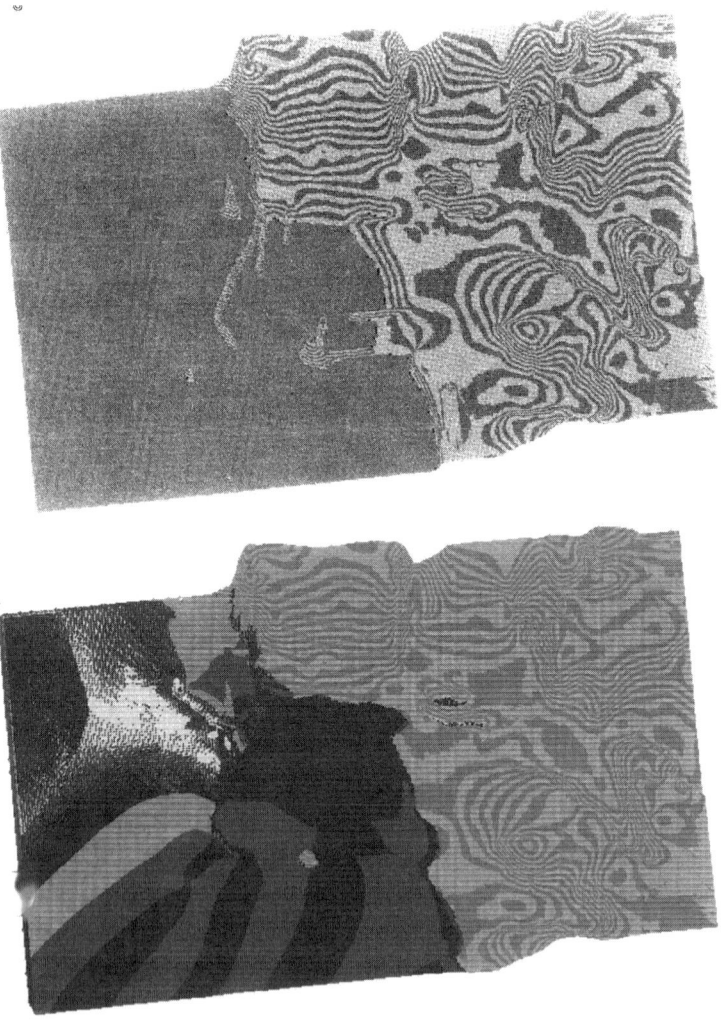

Fig. 2.23. The flooding of Crescent City by the March 28, 1964 tsunami. The contour interval is 0.5 meter.

2G. The 1994 Skagway Tsunami

Introduction

On November 3, 1994, at about 7:10 p.m., a tsunami wave with a period of about 3 minutes and maximum height of 25 to 30 feet occurred in the Taiya Inlet at Skagway, Alaska as reported by Lander[36]. Figure 2.24 is a sketch of Taiya Inlet.

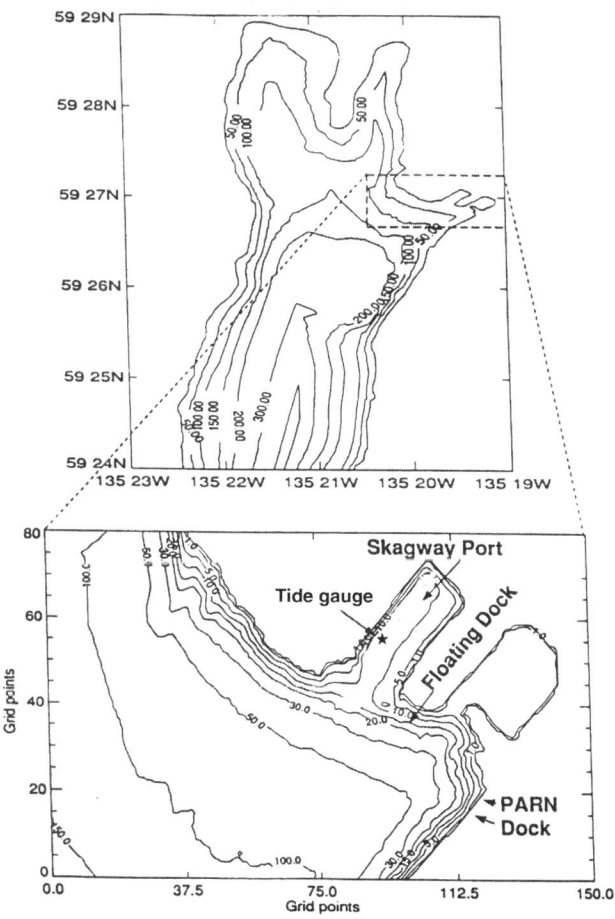

Fig. 2.24. The Taiya Inlet and the Skagway PARN dock and tide gauge location. The depth contours are in meters.

THE SHALLOW WATER MODEL   Chap. 2

The event occurred at the time of a low tide of 4 feet below lower low level. The tidal range at Skagway is about 25 feet. The wave was observed traveling along the Pacific and Arctic Railway and Navigation Company (PARN) dock from the South or deep end of Taiya Inlet. Figure 2.25 is a sketch of the PARN dock.

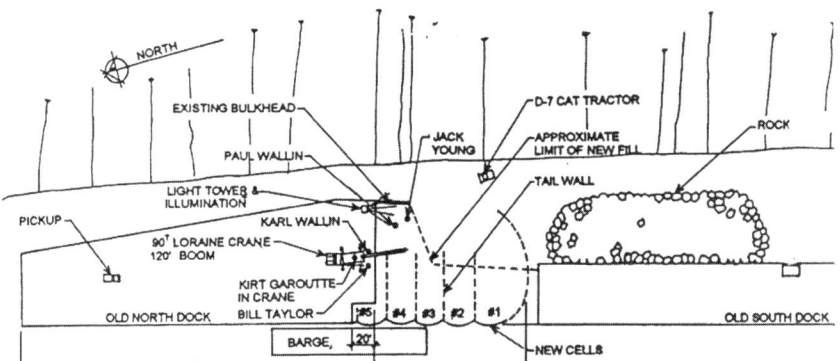

Fig. 2.25. The PARN dock before the landslide generated tsunami occurred.

Kulikov, Rabinovich, Thomson and Bornhold[37] proposed that the tsunami was caused by the collapse of the PARN dock. Their slide involved sediment of 10 to 20 meters thick and extended 125 to 200 meters offshore. Lander[36] estimated the dock slide to have been 600 feet wide, 50 to 60 feet thick, and 4500 feet long with a total volume of 1 to 3 million cubic yards.

From eye witness reports, Bruce Campbell and Dennis Nottingham [38,39,40] have reconstructed the chronology of events.

Time   Event

−2 min   Wind stopped.
0–2 sec   Dock construction cell piles rattled, crane moved.
2–3 sec   Dock and cell piles started to slide seaward.
3–4 sec   Ground fell out from under K. Wallin.
4 sec   K. Wallin was hit with falling wood pile.

Sec. 2G   THE 1994 SKAGWAY TSUNAMI

Time	Event
4–5 sec	Observers saw an incoming wall of water from south end of dock.
4–5 sec	Observers saw wave with south dock, decking and gangway moving toward them.
4–5 sec	Wall of water moving along dock seen before cell piles disappeared.
6 sec	Blue work barge was visible above top of dock.
17–19 sec	Ferry terminal lights disappeared.
18 sec	Loud crash and boom from Ferry Terminal.
120+ sec	Workman returned to crane on North end of dock.
120+ sec	Section of South dock and gangway was at end of North dock.

The NOAA tide gauge was situated midway along the ore dock on the west side of the harbor. The tsunami recorded by this gauge on November 3, 1994 had a period of approximately 3 minutes, a maximum recorded amplitude of about 3 feet and lasted for about 30 minutes. The tide gauge record is shown in Figure 2.26.

The Skagway tide gauge was a damped nitrogen bubbler analog gauge which gives a nonlinear response at short periods. Thus the recorded wave heights are considerably smaller than those of the actual tsunami. To determine the tsunami wave heights, Nottingham in reference 39 determined the water tower generated tide gauge trace that would duplicate the tsunami tide gauge record. The details of the time and wave heights that matched the Skagway tidal chart tsunami are shown in Figure 2.26. The tsunami wave had an initial amplitude of 12 feet and period of 60 sec followed by 3 waves with amplitudes of 6 to 9 feet and period of 180 sec.

Raichlen, Lee, Petroff and Watts[41] also calibrated the Skagway tide gauge. For excitation periods greater than 1000 sec, they found that the gauge gave the wave amplitude. For periods of 3 minutes the gage gave amplitudes that were 40 to 75 percent of the actual wave amplitude. For periods of 20 to 30 sec the gauge gave amplitudes that were 5 to 24 percent of the actual wave amplitude.

THE SHALLOW WATER MODEL   Chap. 2

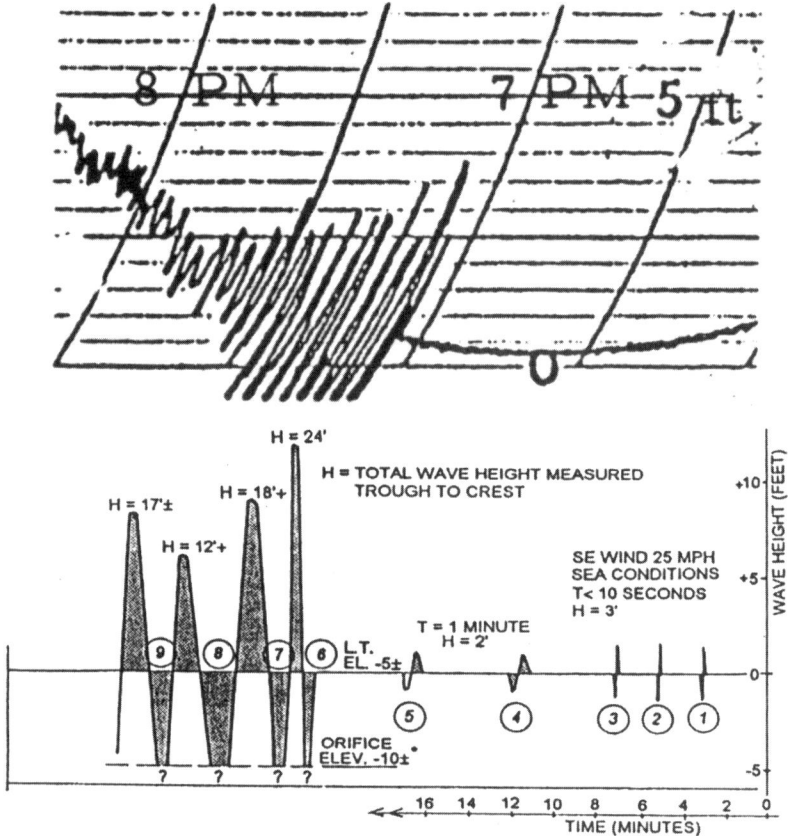

Fig. 2.26. The wave profile that reproduced the Skagway tsunami tide gauge record shown.

Modeling

The generation and propagation of the tsunami wave of November 3, 1994 in the Taiya Inlet was modeled using a 23 by 23 meters grid of the topography. The modeling was performed using the *SWAN* non-linear shallow water code, which includes Coriolis and frictional effects, described in this chapter.

Sec. 2G   THE 1994 SKAGWAY TSUNAMI

A 3 by 6 second land topography was generated from the Rocky Mountain Communication's CD-ROM compilation of the Defense Mapping Agency (DMA) 1 by 1 degree blocks of 3 arc second elevation data. The 23 by 23 meters land topography was generated by interpolation. The sea floor topography before the event was generated by T. Gere of PN&D and Z. Kowalik of the University of Alaska. The grid was 160 by 400 cells and the time step was 0.15 sec.

Dockslide Model

About 750 feet of the dock was destroyed and the slope supporting the dock slid into deep water. The sea floor elevation below the dock area was about 100 feet lower after the event. To model the dock landslide an extreme case was considered. A 740 feet by 1500 feet wide region along the dock was lowered by 100 feet and then the seaward 740 feet by 1500 feet region was raised by 100 feet. The dock slide geometry is shown in Figure 2.27.

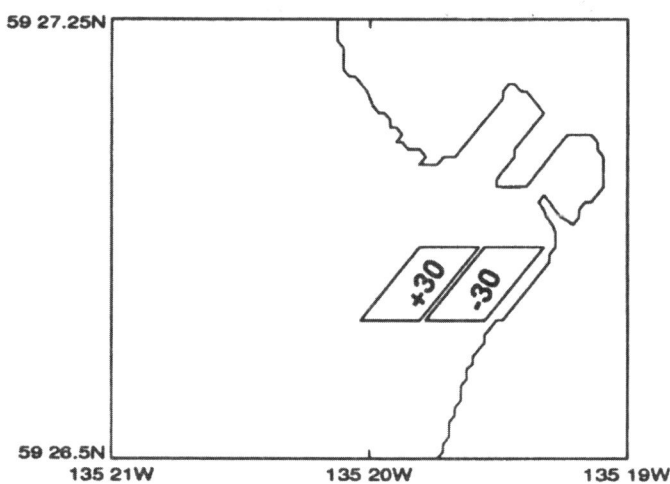

Fig. 2.27. The dockslide model which had dimensions of 450 by 225 by −30 meters drop near the shoreline and 450 by 225 by 30 meters uplift seaward.

81

THE SHALLOW WATER MODEL Chap. 2

The calculated wave profile at the south end of the PARN dock (cell (143,227) at 28 meters depth) and the wave profile at the tide gauge (cell (140,245) at 13 meters depth) are shown in Figure 2.28.

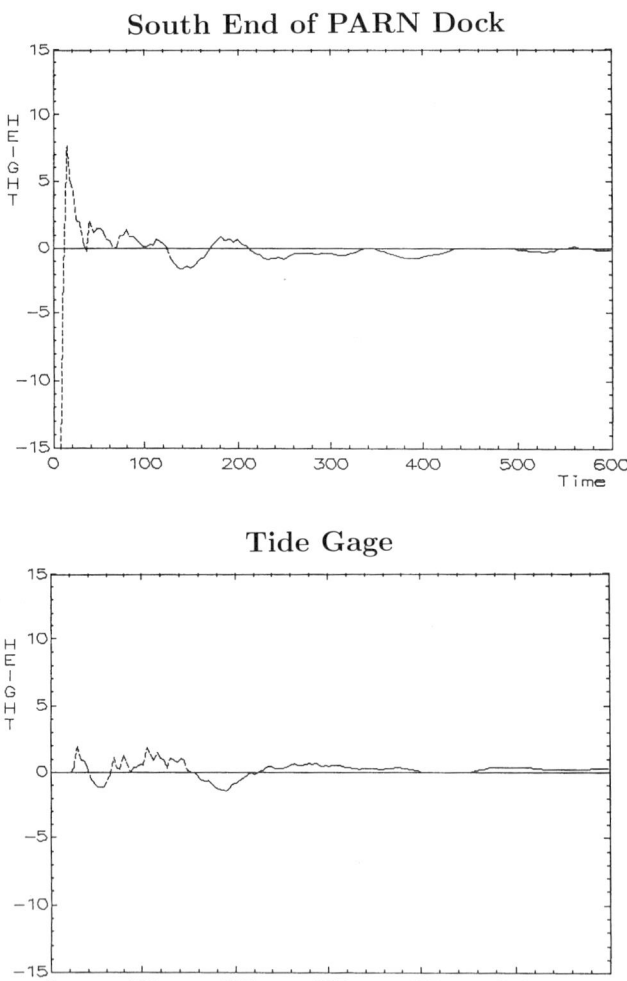

Fig. 2.28. The wave profiles at the South end of the PARN dock and at the tide gauge are shown for the dockside model. The vertical axis is height in meters and the time is in sec.

Sec. 2G   THE 1994 SKAGWAY TSUNAMI

The dockslide landslide generated a tsunami wave with much shorter wave periods than observed (less than a third). The direction of the wave was 90 degrees different than observed. Since the dimensions of the slide used in the model were as large as possible from the surveys, the dock landslide could not have generated the observed tsunami wave.

Inlet Landslide Model

The sea floor elevations before and after the event indicate that a considerable area of the sea floor was lower for about 5000 feet down the inlet and then much of the sea floor was higher further down the inlet as described by Campbell in references 38, 39 and 40. NOAA performed an extensive underwater mapping that accurately located the landslides that caused the 1994 tsunami. They are shown in Figures 2.29 and 2.30.

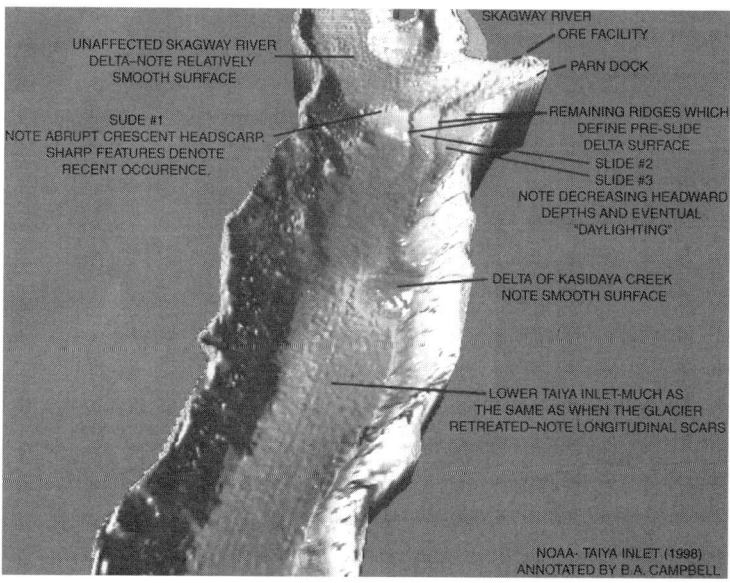

Fig. 2.29. The NOAA underwater map of Taiya Inlet. The subsea slides are clearly visible. Figure 2.30 zooms into the region of the harbor.

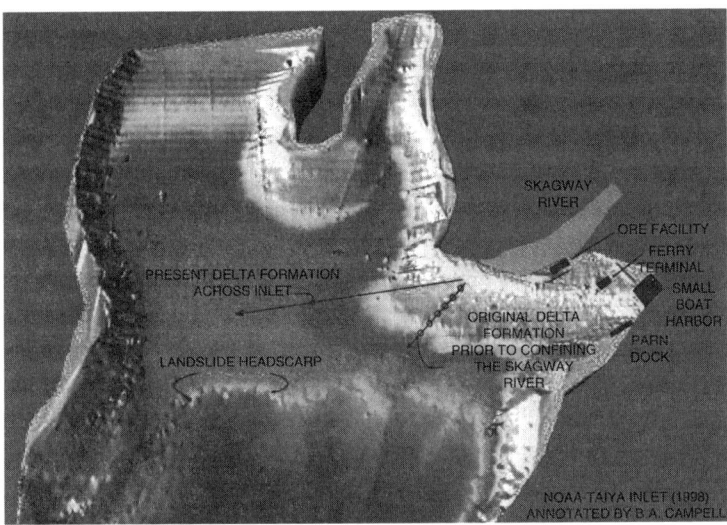

Fig. 2.30. An expanded view of Figure 2.29 showing Skagway harbor and the multiple subsea slide regions. The directions of the Skagway River sediment movement currently and historically before the river was confined are shown.

The sea floor elevation changes had a volume of 20 to 25 million cubic yards. The landslide model studied was that described by Bruce Campbell in references 38 and 39 from the change in the sea floor topography that occurred before and after the tsunami. The seafloor before and after the landslide is shown in Figures 2.31 and 2.32. The Campbell model consists of three slide regions separated by ridges. The west slide region was approximately 1000 feet wide, 4000 feet long and had a volume of 11.4 million cubic yards. The middle slide region was 500 feet wide, 3600 feet long and had a volume of 5.1 million cubic yards. The east slide region was 400 feet wide, 4800 feet long and had a volume of 5 million cubic yards.

The material from the three slides moved down into the deep part of the inlet occupying an area 3000 feet wide by 5800 feet long with a volume of 21.5 million cubic yards.

The profiles of the slide depths showed that the upper region of each slide was thicker than the rest of the slide. The slide regions were described by a thicker upper region followed by a less thick lower region.

Sec. 2G THE 1994 SKAGWAY TSUNAMI

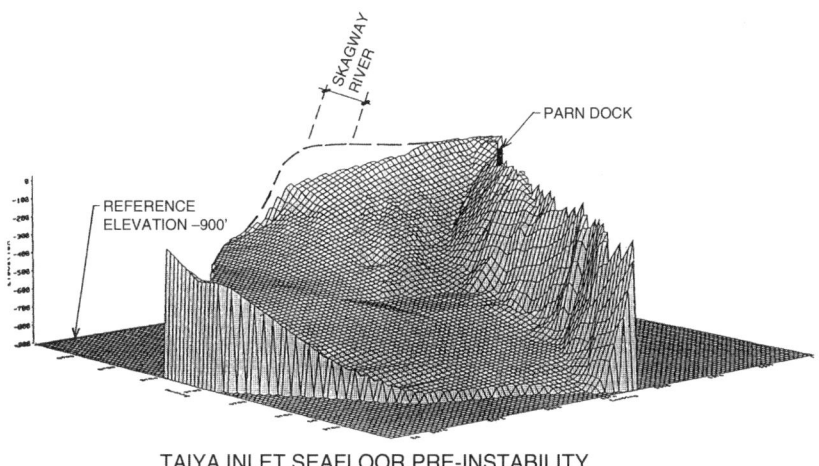

Fig. 2.31. Taiya Inlet topography before the 1994 tsunami.

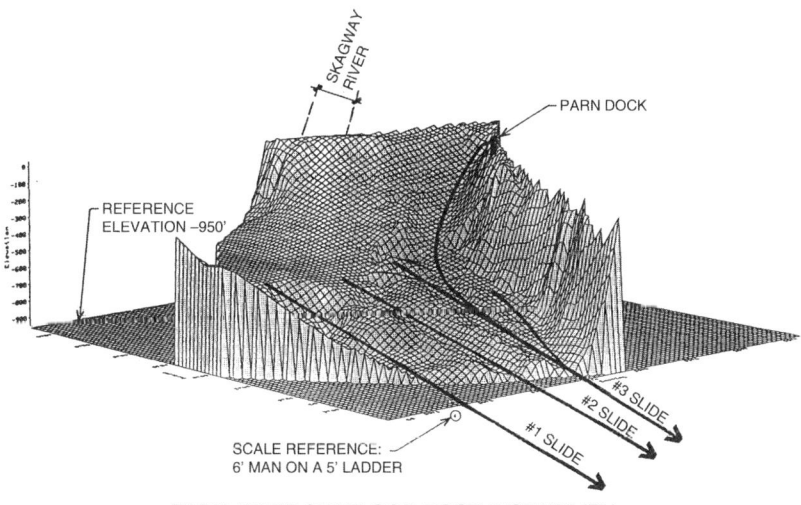

Fig. 2.32. Taiya Inlet topography after the 1994 tsunami. The three slides, the PARN dock and the outlet of the Skagway River that was the source of the sediment that caused the 1994 tsunami.

THE SHALLOW WATER MODEL   Chap. 2

The Campbell 3-slide model is shown in Figure 2.33 and was described numerically as follows:

West Slide –upper region– 1056 by 1056 by 137.8 feet deep.
West Slide –lower region– 1056 wide by 2942 long by 49.2 feet deep.
West Slide – Total Volume of 11.4 million cubic yards.

Middle Slide -upper region- 528 wide by 1207 long by 118.8 feet deep.
Middle Slide -lower region- 528 wide by 2414 long by 49.2 feet deep.
Middle Slide - Total Volume of 5.11 million cubic yards.

East Slide –upper region– 377 wide by 1131 long by 98.4 feet deep.
East Slide –lower region– 377 wide by 3696 long by 65.6 feet deep.
East Slide – Total Volume of 4.94 million cubic yards.

Slide Debris – 3017 wide by 5809 long by 32.8 feet thick.
Slide Debris – Total Volume of 21.3 million cubic yards.

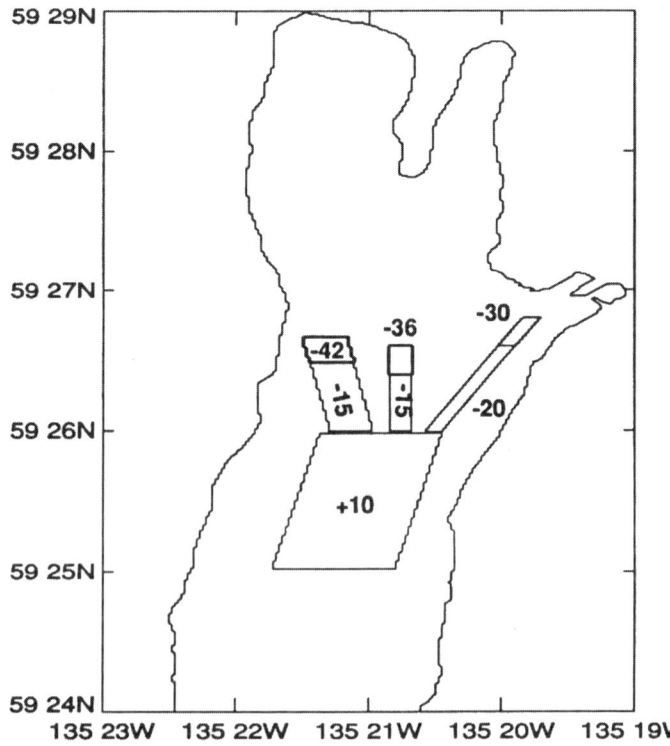

Fig. 2.33. The Campbell 3-slide Landslide model. The dimensions are in meters.

Sec. 2G   THE 1994 SKAGWAY TSUNAMI

The calculated wave profile at the south end of the PARN dock (cell (143,227) at 28 meters depth) and the wave profile at the tide gauge (cell (140,245) at 13 meters depth) are shown in Figure 2.34 for the landslide displacement occurring instantaneously.

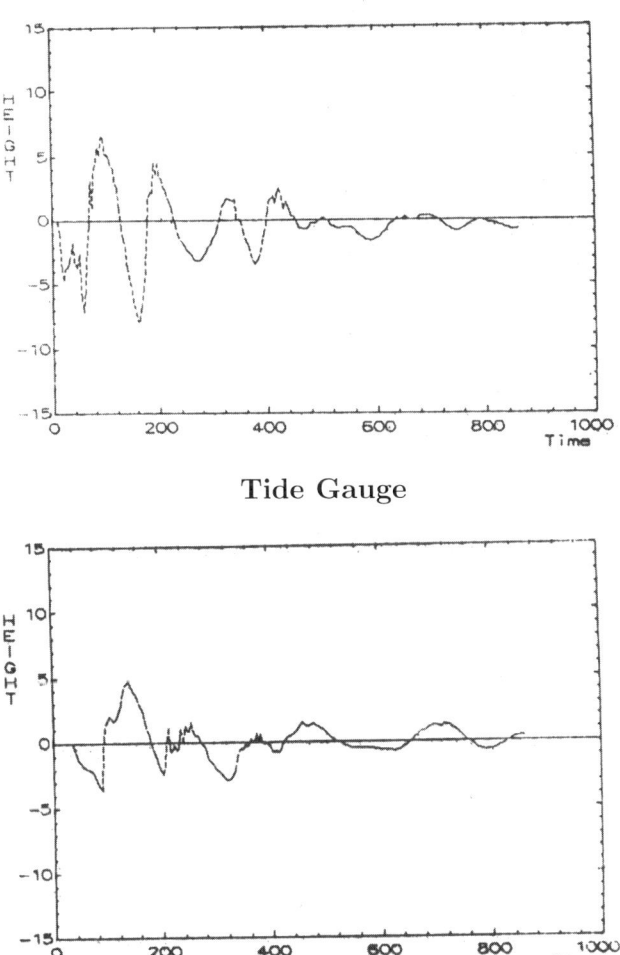

Fig. 2.34. The wave profiles at the South end of the PARN dock and at the tide gauge are shown for the 3-slide Landslide model. The vertical axis is height in meters and the time is in sec.

THE SHALLOW WATER MODEL   Chap. 2

The time for the landslide to occur was varied from 0 to 3 minutes. The calculated wave period difference for the landslide and dockslide was insensitive to the time for the landslide to occur and also to whether the upper half of the landslide occurred as 3 slides or as a single slide.

The tide gauge duration and profile were best reproduced by the 2 minute duration slide (about 26.5 meters/sec). The calculated and calibrated tide gauge profiles are compared in Figure 2.35.

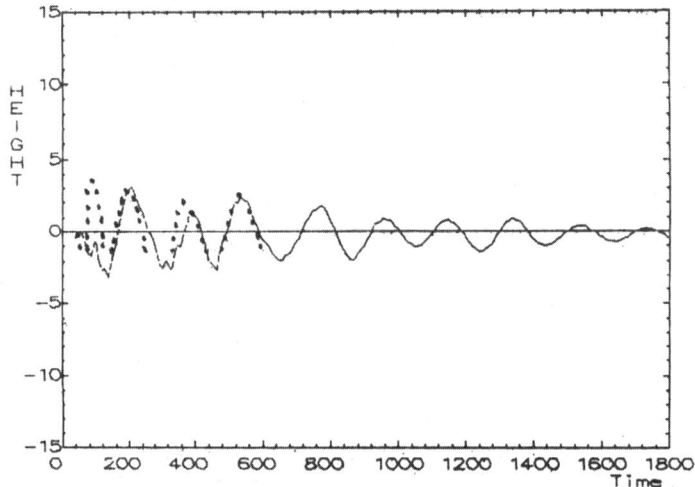

Fig. 2.35. The wave profile at the tide gauge generated by the landslide occurring over a 2 minute interval and the calibrated tide gauge profile (dots). The vertical axis is height in meters and the time is in sec.

The calculations using the *SWAN* code are described in reference 42. The computer movies are on the NMWW CD-ROM in the directory /TSUNAMI.MVE/SKAGWAY.

Professor Kowalik of the University of Alaska reported his numerical modeling study of the Skagway landslide generated tsunami in reference 43. He used the Campbell 3-slide Landslide model and obtained the wave profile shown in Figure 2.36 at the tide gauge location.

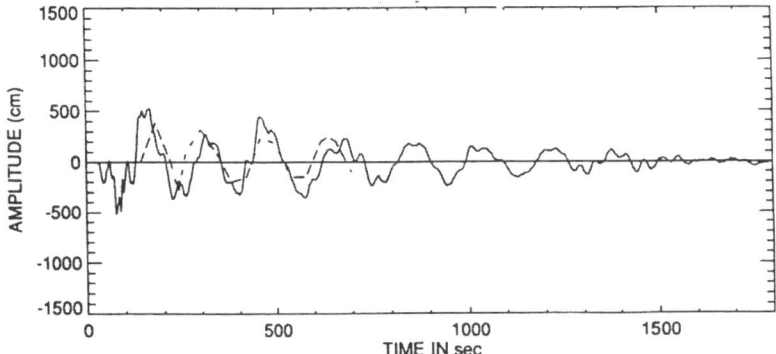

Fig. 2.36. Tide gauge profile from calculations performed by Professor Zygmunt Kowalik of the University of Alaska. The wave profile at the tide gauge was generated by the landslide occurring with a 32.7 meters/sec velocity and the calibrated tide gauge profile (dashes). The vertical axis is height in centimeters and the time is in sec.

Again, the computed tide gauge profile was in satisfactory agreement with the observations except for the initial 50 sec wave which was not included in the observed profile in Figure 2.36.

The remarkably long observed and calculated 30 minute duration of the tide gage record is a result of the resonance for 3 minute waves in the harbor where the tide gauge is located. The waves are decayed outside the harbor after 10 minutes. A similar harbor resonance was described in reference 41 and studied in detail by Kowalik in reference 43.

Kowalik[43] also studied the motion of the floating dock. The floating dock was located in the harbor at the end of the jetty as shown in Figure 2.24. It serves as the Alaskan Ferry Terminal. The tsunami caused about one million dollars damage to the floating dock, but replacement of the dock cost over twenty million dollars. The floating dock was moored with thirteen chains, thus allowing movement of the dock with the large tides. During the tsunami event all chains were broken and the dock moved violently west and then east from the original location.

The complicated movement of the dock during the tsunami event was described by Raichlen, et al.[41]. Kowalik's modeling showed the same strong motions of the floating dock to the east and to the west from the initial locations as was observed.

The initial tide gauge 50 sec peak was not reproduced by any of the models. Nottingham[44] performed physical modeling in a chamber, whose pressure could be varied, of the behavior of the type of tide gauge that was at Skagway during the 1994 tsunami event. Campbell and Nottingham [38,39,40] reported that southerly winds around 25 miles/hour were blowing on November 3, 1994, but suddenly stopped about one or two minutes before the tsunami arrived at the PARN dock. Obviously a change in barometric pressure occurred. Nottingham evaluated the effect of barometric pressure change on the Bubbler tide gauge with the gauge and orifice completely enclosed in a pressure/vacuum chamber.

A negative air pressure of about two inches of Mercury would cause an immediate and rapid gauge record movement similar to the initial trace recorded on the gauge shown in Figure 2.26. Superposition of the trace produced during the vacuum test and the trace from the dampened standpipe test produced a trace almost exactly as recorded by the tide gauge at Skagway on November 3, 1994.

When the landslide occurred, the water surface above the landslide fell rapidly, creating a decrease in air pressure which resulted in the observed wind stopping at Skagway and the initial 50 sec peak in the tide gauge record.

Three-Dimensional full Navier-Stokes Modeling

In Chapter 4, the technique for modeling three-dimensional, full Navier-Stokes, nonlinear, incompressible viscous water waves and the code *SOLA-3D* is described. While the wavelengths of the tsunamis generated by the Skagway landslides were long compared to the depth, the limitations of the shallow water model described in Chapter 5 were a concern during the study of the 1994 Skagway tsunami. A single slide model, with the same mass as the 3-slide model, for the Skagway tsunami that gave about the observed period was used for the study.

A three-dimensional *SOLA-3D* calculation and a *SWAN* calculation for the same single slide model and with similar resolution was performed. The computer movies are on the NMWW CD-ROM in the directory /TSUNAMI.MVE/SOLA/SKAGWAY. The wave profiles generated by the *SOLA-3D* code and the *SWAN* code were similar in amplitude and periods indicating that the extensive studies of the 1994 Skagway tsunami were not seriously flawed by the shallow water assumption.

Conclusions

The dock landslide generated a calculated tsunami wave with much shorter wave periods than observed (less than a third). The direction of the wave was 90 degrees different than observed. Since the dimensions of the slide used in the model were about as large as possible from the surveys, the dock landslide could not have generated the observed tsunami wave.

The complicated bottom topography changes that occurred were described in the numerical model using a 3-slide region model. The landslides generated a tsunami wave with about the wave amplitude and period observed. The direction of the wave was the same as observed along the dock.

As discussed by Nottingham in reference 45, the Skagway River continues to move massive amounts of sediment out beyond Skagway harbor and it is only a matter of time until another landslide occurs and generates a tsunami. If the tsunami should occur at high tide rather than low tide, the town of Skagway is at serious risk. If the tsunami should occur when two or three cruise ships were docked at the harbor, the loss of lives and the damage to the ships could be very large.

A plan[45], under the National Tsunami Hazard Mitigation Program, is being developed to create warning and mitigation measures for Skagway. Mitigation measures and a warning system for locally generated underwater landslides would be useful to other parts of the world with similar problems.

References

1. W. Hansen, "Theorie zur Errechnung des Wasserstandes und der Strömungen in Tandmeeren nebst Anwendungen," Tellus, Vol. 8, No. 3 (1956).
2. J. Proudman, *Dynamical Oceanography*, John Wiley & Sons, Inc., New York (1953).
3. J. J. Dronkers, *Tidal Computations in Rivers and Coastal Waters*, John Wiley & Sons, Inc., New York (1964).
4. P. Welander, "Numerical Prediction of Storm Surges," *Advances in Geophysics*, Vol. 8, H. E. Landsberg and J. van Miegham (Eds.), Academic Press, New York (1961).
5. W. Hansen, "Hydrodynamical Methods Applied to Oceanographic Problems," Proceedings of the Symposium on Mathematical Hydrodynamic Methods of Physical Oceanography, Institut für Meereskunde der Universität, Hamburg (1962).
6. S. Uusitalo, "The Numerical Calculation of Wind Effect in Sea-level Elevations," Proceedings of the Symposium on Mathematical Hydrodynamic Methods of Physical Oceanography, Institut für Meereskunde der Universität, Hamburg (1962).
7. Jan J. Leenderstse, "Aspects of a Computational Model for Long Period Water Wave Propagation," Rand Corporation report RM–5294–PR (1967).
8. H. G. Ramming and Z. Kowalik, *Numerical Modeling of Marine Hydrodynamics*, Elsevier Scientific Publishing Co., New York (1980).
9. T. S. Murty, "Seismic Sea Waves – Tsunamis," Fisheries and Marine Service Department of Fisheries and the Environment, Ottawa, Canada, Bulletin 198 (1977).
10. T. S. Murty, "Storm Surges – Meteorological Ocean Tides," Fisheries and Marine Service Department of Fisheries and the Environment, Sidney, British Columbia, Bulletin 212 (1984).
11. An. G. Marchuk, L. B. Chubarov, and Iu. I. Shokin, *Numerical Modeling of Tsunami Waves*, Nauka Press, Siberian Branch, Novosibirsk (1983). Available in English as LA–TR–85–40 to U. S. government agencies and their contractors and from Leo Kanner Associates, Redwood City, California.

REFERENCES

12. E. N. Pelinovsky, "Tsunami Climbing A Beach," Institute of Applied Physics, U.S.S.R. Academy of Science, Gorky (1985).
13. Charles L. Mader, "Numerical Simulation of Tsunamis," Journal of Physical Oceanography, Vol. 4, pp. 74–82 (1974).
14. Safwan Hadi, "A Numerical Tidal Model of Musi-Upang Estuaries," a dissertation submitted to Oceanography Department of University of Hawaii (1985).
15. Charles L. Mader and Sharon Lukas, "Numerical Modeling of Waianae Harbor," Aha Hulikoa Hawaiian Winter Workshop Proceedings (January 1985).
16. Charles L. Mader, Martin Vitousek, and Sharon Lukas, "Numerical Modeling of Atoll Reef Harbors," Proceedings of the International Symposium on Natural and Man-Made Hazards, Rimouski (1986).
17. Charles L. Mader and Sharon Lukas, "*SWAN* – A Shallow Water, Long Wave Code: Applications to Tsunami Models," Joint Institute for Marine and Atmospheric Research report JIMAR 84–077 (1984).
18. Charles L. Mader and George Curtis, "Modeling Hilo, Hawaii Tsunami Inundation," Science of Tsunami Hazards, Vol. 9, pp. 85–94 (1991).
19. Charles L. Mader and George Curtis, "Numerical Modeling of Tsunami Inundation of Hilo Harbor," University of Hawaii JIMAR Contribution No. 91-251 (1991).
20. Charles L. Mader, "Modeling of Tsunami Waves," Scientific Computing and Automation, June Issue, pp. 19–23 (1993).
21. C. K. Green, "Seismic Sea Wave of April 1, 1946 as Recorded on Tide Gauges," Transactions of the American Geophysical Union, Vol. 27, pp. 490–502 (1946).
22. A. S. Furumoto, private communication (1991).
23. Hilo Harbor Model Conference on 23–24 November 1964 at Look Laboratory of Oceanographic Engineering, Honolulu, Hawaii, Corps of Engineers, U.S. Army Engineer District, Honolulu.
24. "Advanced Information for Participants Hilo Harbor Model Conference on 23–24 November 1964 at Look Laboratory of Oceanography, Honolulu" by U.S. Army Engineering District, Honolulu, Hawaii.

25. "Physically Feasible Means for Protecting Hilo from Tsunamis," Third Report of the Hilo Technical Tsunami Advisory Council to the Board of Supervisors, Hawaii County through its Tsunami Advisory Committee, December 31, 1965. The committee was Doak C. Cox, Masashi Hom-ma, Masatsugu Suzuki, Ryutaro Takahasi and Robert L. Wiegel.
26. William G. Van Dorn, "Tsunami Response at Wake Island," Journal of Marine Research, Vol. 28, pp. 336–344 (1970).
27. G. Plafker, "Tectonics of the March 27, 1964 Alaska Earthquake" U.S. Geological Survey Professional Paper 543-I, I1-I74 (1969).
28. Li-San Hwang and D. Divoky, "Numerical Investigations of Tsunami Behavior," Tetra Tech, Inc. report (1975).
29. "A Numerical Model of the Major Tsunami," THE GREAT ALASKA EARTHQUAKE OF 1964, National Academy of Sciences (1972).
30. James R. Houston, Robert W. Whalin, Andrew W. Garcia and H. Lee Butler, "Effect of Source Orientation and Location in the Aleutian Trench on Tsunami Amplitude along the Pacific Coast of the Continental United States," Research Report H-75-4 of U.S. Army Engineering Waterways Experiment Station, Vicksburg, Mississippi.
31. G. Plafker and J. C. Savage, "Mechanism of the Chilean Earthquakes of May 21 and 22, 1960," Geological Society of America Bulletin, Vol. 81, pp. 1001–1030 (1970).
32. Zygmunt Kowalik and Paul M. Whitmore, "An Investigation of Two Tsunamis Recorded at Adak, Alaska," Science of Tsunami Hazards, Vol. 9, pp. 67–83 (1991).
33. Paul M. Whitmore and Thomas J. Sokolowski, "Predicting Tsunami Amplitudes along the North Amerian Coast from Tsunamis Generated in the Northwest Pacific Ocean During Tsunami Warnings," Science of Tsunami Hazards, Vol. 14, pp. 147–166 (1996).
34. Charles L. Mader, George D. Curtis and George Nabeshima, "Modeling Tsunami Flooding of Hilo, Hawaii," Recent Advances in Marine Science and Technology, 92, pp. 79–86 PACON International (1993).

REFERENCES

35. E. Bernard, C. Mader, G. Curtis, and K. Satake, "Tsunami Inundation Model Study of Eureka and Crescent City," NOAA Technical Memorandum ERL PMEL-103 (1994).
36. James F. Lander, "Tsunamis Affecting Alaska 1737–1996," NGDC Key to Geophysical Research Documentation No. 31 (1996).
37. E. A. Kulikov, A. B. Rabinovich, R. E. Thomson and B. D. Bornhold, "The Landslide Tsunami of November 3, 1994, Skagway Harbor, Alaska," Journal of Geophysical Research, Vol. 101, pp. 6609–6615 (1996).
38. Bruce Campbell, "Report of a Sea floor Instability at Skagway, Alaska – November 3, 1994," Campbell and Associates Report of January 16, 1995.
39. Bruce Campbell, "Skagway Seafloor Instability Re-Analysis and Update," Campbell and Associates Report of January 28, 1997.
40. Bruce A. Campbell and Dennis Nottingham, "Anatomy of a Landslide Created Tsunami at Skagway, Alaska November 3, 1994," Science of Tsunami Hazards, Vol. 17, pp. 19–48 (1999).
41. Fredric Raichlen, Jiin Jen Lee, Catherine Petroff, and Philip Watts, "The Generation of Waves by a Landslide: Skagway, Alaska – A Case Study," Proceedings of 25th International Conference on Coastal Engineering (1996).
42. Charles L. Mader, "Modeling the 1994 Skagway Tsunami," Science of Tsunami Hazards, Vol. 14, pp. 18–48 (1996).
43. Zygmunt Kowalik, "Landslide-Generated Tsunami in Skagway, Alaska," Science of Tsunami Hazards, Vol. 15, pp. 67–80 (1997).
44. Dennis Nottingham, "The 1994 Skagway Tsunami Tide Gage Record," Science of Tsunami Hazards, Vol. 15, pp. 81–106 (1997).
45. Dennis Nottingham, "Review of the 1994 Skagway, Alaska Tsunami and Future Plans," Science of Tsunami Hazards, Vol. 20, pp. 42–49 (2002).

3

THE TWO-DIMENSIONAL NAVIER-STOKES MODEL

3A. The Two-Dimensional Navier-Stokes Equations

The two-dimensional time dependent Navier-Stokes equations may be derived from equations 1.4 and 1.7, setting all terms in z to zero.

$$\frac{\partial U_x}{\partial x} + \frac{\partial U_y}{\partial y} = 0 \ . \tag{3.1}$$

$$\frac{\partial U_x}{\partial t} + U_x \left(\frac{\partial U_x}{\partial x}\right) + U_y \left(\frac{\partial U_x}{\partial y}\right)$$
$$= -\frac{1}{\rho_o}\frac{\partial P}{\partial x} + g_x + \frac{\mu}{\rho_o}\left(\frac{\partial^2 U_x}{\partial x^2} + \frac{\partial^2 U_x}{\partial y^2}\right) \ . \tag{3.2}$$

$$\frac{\partial U_y}{\partial t} + U_x \left(\frac{\partial U_y}{\partial x}\right) + U_y \left(\frac{\partial U_y}{\partial y}\right)$$
$$= -\frac{1}{\rho_o}\frac{\partial P}{\partial y} + g_y + \frac{\mu}{\rho_o}\left(\frac{\partial^2 U_y}{\partial x^2} + \frac{\partial^2 U_y}{\partial y^2}\right) \ . \tag{3.3}$$

If we wish to replace the x direction with a radius r, we introduce a geometry term α, where $\alpha = 0$ for plane (Cartesian) coordinates and $\alpha = 1.0$ for cylindrical coordinates. Equations 3.1–3.3 become, after some manipulation and substitution,

THE TWO-DIMENSIONAL NAVIER-STOKES MODEL Chap. 3

$$\frac{1}{r^\alpha}\frac{\partial r^\alpha U_r}{\partial r} + \frac{\partial U_y}{\partial y} = 0 \, . \tag{3.4}$$

$$\frac{\partial U_r}{\partial t} + \frac{1}{r^\alpha}\frac{\partial r^\alpha U_r^2}{\partial r} + \frac{\partial U_r U_y}{\partial y}$$

$$= -\frac{1}{\rho_o}\frac{\partial P}{\partial r} + g_r + \frac{\mu}{\rho_o}\frac{\partial}{\partial y}\left(\frac{\partial U_r}{\partial y} - \frac{\partial U_y}{\partial r}\right) \, .$$

$$\frac{\partial U_y}{\partial t} + \frac{1}{r^\alpha}\frac{\partial r^\alpha U_r U_y}{\partial r} + \frac{\partial U_y^2}{\partial y}$$

$$= -\frac{1}{\rho_o}\frac{\partial P}{\partial y} + g_y - \frac{\mu}{\rho_o}\frac{1}{r^\alpha}\frac{\partial}{\partial r}\left[r^\alpha\left(\frac{\partial U_r}{\partial y} - \frac{\partial U_y}{\partial r}\right)\right] \, .$$

The conversion to radial geometry is described in Appendix D of reference 7. The term

$$\frac{1}{r^\alpha}\frac{\partial r^\alpha U_r^2}{\partial r} + \frac{\partial U_r U_y}{\partial y} \tag{3.5}$$

is equal to the more usual expression

$$U_r\frac{\partial U_r}{\partial r} + U_y\frac{\partial U_r}{\partial y} \, , \tag{3.6}$$

after substitution of the mass equation 3.4 and some manipulation.

Sec. 3B THE TWO-DIMENSIONAL NAVIER-STOKES EQUATIONS

The coding equations used in the *ZUNI* code also set $\frac{P}{\rho_o} = \phi$, $U_x = u$, $U_y = v$, $y = z$, and $\frac{\mu}{\rho_o} = \nu$ to get the following set of curious basic differential equations used in the *ZUNI* code

$$D \equiv \frac{1}{r^\alpha} \frac{\partial r^\alpha u}{\partial r} + \frac{\partial v}{\partial z} = 0 , \tag{3.7}$$

$$\frac{\partial u}{\partial t} + \frac{1}{r^\alpha} \frac{\partial r^\alpha u^2}{\partial r} + \frac{\partial uv}{\partial z}$$
$$= -\frac{\partial \phi}{\partial r} + g_r + \nu \frac{\partial}{\partial z} \left(\frac{\partial u}{\partial z} - \frac{\partial v}{\partial r} \right) , \text{ and} \tag{3.8}$$

$$\frac{\partial v}{\partial t} + \frac{1}{r^\alpha} \frac{\partial r^\alpha uv}{\partial r} + \frac{\partial v^2}{\partial z}$$
$$= -\frac{\partial \phi}{\partial z} + g_z - \frac{\nu}{r^\alpha} \frac{\partial}{\partial r} \left[r^\alpha \left(\frac{\partial u}{\partial z} - \frac{\partial v}{\partial r} \right) \right] , \tag{3.9}$$

where D is a "discrepancy term" used as described later.

3B. The Finite-Difference Equations

The finite-difference technique used for solving the time dependent Navier-Stokes equations (3.1–3.3) is the Marker and Cell (MAC) method of Harlow and Welch.[1] It is a numerical technique for calculating viscous, incompressible flow with a free surface.

THE TWO-DIMENSIONAL NAVIER-STOKES MODEL

The MAC method is based on an Eulerian network of rectangular cells, with velocities centered at cell boundaries and the pressure cell centered. Just as the differential equations of motion are statements of the conservation of mass and momentum, the MAC finite-difference equations express these conservation principles for each cell, or combination of cells, in the computing mesh.

After the introduction of MAC, much attention was given to devising more accurate treatments of the free surface boundary conditions. Hirt and Shannon[2] incorporated improved surface approximations to the normal stress conditions and showed the necessity for satisfying the tangential stress conditions. Chan and Street[3] developed a technique for more accurate delineation of the free surface. This permitted the free surface pressure to be specified at the surface itself, rather than at the center of the surface cell. Nichols and Hirt[4] modified the Chan and Street procedure, devising a technique for defining the fluid surface by a set of surface marker particles that move with local fluid velocity. These particles allow surface cell pressures in MAC to be accurately specified by means of linear interpolation or extrapolation between the known values of pressure in the nearest full cell and the desired fluid surface.

A Simplified MAC (SMAC), described by Harlow and Amsden,[5] does not require the pressure to be calculated. It is possible to use the actual pressures or any initial pseudopressure (for internal cells) such as zero.

The computing program is called *ZUNI* and is described by Amsden.[6] Parts of Amsden's report are included in the following description of the *ZUNI* code. The SMAC technique of Harlow and Amsden has been modified to include the Nichols and Hirt free surface treatment. A partial cell treatment that allows a rigid free slip obstacle to be placed through cell diagonals has also been included. The desired boundary slope is obtained by choosing the appropriate aspect ratio for the cells of the mesh. Thus, the numerical technique can be used to calculate wave runup on exposed beaches in addition to submerged beaches. The procedure for a calculational cycle is as follows:

Sec. 3B   THE FINITE-DIFFERENCE EQUATIONS

1) A tentative field of advanced time velocities is calculated by using an arbitrary pressure field within the fluid, but with a pressure boundary condition at the free surface satisfying the normal stress condition. Correct velocity boundary conditions ensure that this tentative velocity field contains the correct vorticity at every interior point in the fluid. The tentative velocities do not, however, have the conservation of mass equation satisfied.

2) The tentative velocities are modified to their final values so as to preserve the vorticity at every point. A potential function is employed, determined by the requirement that it convert the velocity field to one that satisfies the incompressibility condition everywhere.

In differential form, the equation for transport of vorticity ω is independent of the pressure, so that any field of pressure inserted into the Navier-Stokes equations will ensure that the resulting velocity field carries the correct vorticity. An arbitrary pressure field will not, however, ensure that D vanishes, but if the velocity field is altered by the addition of the gradient of an appropriate potential function, the resulting field will carry the same vorticity, will have vanishing D, and accordingly will be uniquely determined, hence correct.

This is the procedure that is used for the finite-difference approximations to equations 3.7–3.9.

$$D_{i,j}^{n+1} \equiv \frac{r_{i+1/2}^{\alpha}\, \tilde{u}_{i+1/2,j}^{n+1} - r_{i-1/2}^{\alpha}\, \tilde{u}_{i-1/2,j}^{n+1}}{r_i^{\alpha}\, \delta r}$$

$$+ \frac{\tilde{v}_{i,j+1/2}^{n+1} - \tilde{v}_{i,j-1/2}^{n+1}}{\delta z} = 0 \, . \tag{3.10}$$

The subscripts refer to position in the finite-difference mesh, as shown in Figure 3.1, and the superscript n counts time cycles. The true pressure ϕ has been replaced by the arbitrary field θ and, accordingly, the new time velocities are marked with tildes.

THE TWO-DIMENSIONAL NAVIER-STOKES MODEL

$$\frac{\tilde{u}^{n+1}_{i+1/2,j} - u^n_{i+1/2,j}}{\delta t}$$

$$= \frac{r^\alpha_i u^n_{i+1/2,j}\ u^n_{i-1/2,j} - r^\alpha_{i+1}\ u^n_{i+3/2,j}\ u^n_{i+1/2,j}}{r^\alpha_{i+1/2}\delta r}$$

$$+ \frac{u^n_{i+1/2,j-1/2}\ v^n_{i+1/2,j-1/2} - u^n_{i+1/2,j+1/2}\ v^n_{i+1/2,j+1/2}}{\delta z}$$

$$+ \frac{\theta_{i,j} - \theta_{i+1,j}}{\delta r} + g_r$$

$$+ \nu \left[\frac{1}{\delta z^2}\left(u^n_{i+1/2,j+1} + u^n_{i+1/2,j-1} - 2u^n_{i+1/2,j}\right) \right.$$

$$- \frac{1}{\delta r \delta z}\left(v^n_{i+1,j+1/2} - v^n_{i+1,j-1/2} - v^n_{i,j+1/2}\right.$$

$$\left.\left. + v^n_{i,j-1/2}\right)\right]. \tag{3.11}$$

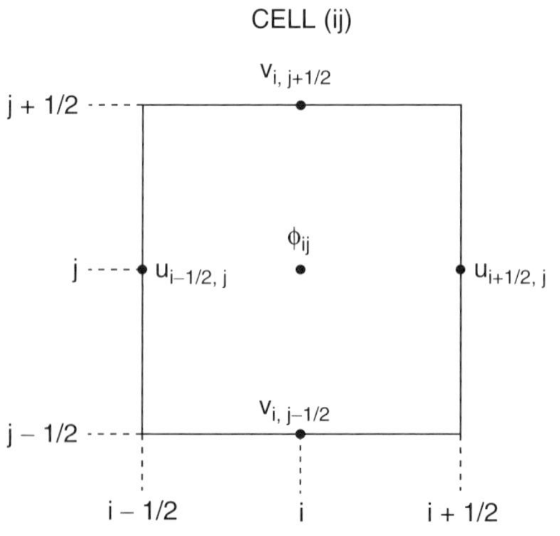

Fig. 3.1. The location of the cell variables in a SMAC cell.

Sec. 3B   THE FINITE-DIFFERENCE EQUATIONS

$$\frac{\tilde{v}_{i,j+1/2}^{n+1} - v_{i,j+1/2}^{n}}{\delta t}$$

$$= \frac{r_{i-1/2}^{\alpha}\ u_{i-1/2,j+1/2}^{n}\ v_{i-1/2,j+1/2}^{n}}{r_{i}^{\alpha}\delta r}$$

$$- \frac{r_{i+1/2}^{\alpha}\ u_{i+1/2,j+1/2}^{n}\ v_{i+1/2,j+1/2}^{n}}{r_{i}^{\alpha}\delta r}$$

$$+ \frac{v_{i,j+1/2}^{n}\ v_{i,j-1/2}^{n} - v_{i,j+3/2}^{n}\ v_{i,j+1/2}^{n}}{\delta z}$$

$$+ \frac{\theta_{i,j} - \theta_{i,j+1}}{\delta z} + g_z$$

$$- \frac{\nu}{r_{i}^{\alpha}\delta r}\left[r_{i+1/2}^{\alpha}\left(\frac{u_{i+1/2,j+1}^{n} - u_{i+1/2,j}^{n}}{\delta z} \right.\right.$$

$$\left. - \frac{v_{i+1,j+1/2}^{n} - v_{i,j+1/2}^{n}}{\delta r} \right)$$

$$- r_{i-1/2}^{\alpha}\left(\frac{u_{i-1/2,j+1}^{n} - u_{i-1/2,j}^{n}}{\delta z} \right.$$

$$\left.\left. - \frac{v_{i,j+1/2}^{n} - v_{i-1,j+1/2}^{n}}{\delta r} \right) \right]. \quad (3.12)$$

The finite-difference equations have been written with explicit (retarded-time) fluxes. Cell-centered momentum convection terms have been written in the ZIP form,[7] which, although continuing to ensure internal momentum conservation, allows SMAC to conserve momentum in the immediate vicinity of a rigid wall. This type of differencing introduces the definition

THE TWO-DIMENSIONAL NAVIER-STOKES MODEL Chap. 3

$$(u_{i,j})^2 \equiv u_{i-1/2,j}\, u_{i+1/2,j} \,. \tag{3.13}$$

An advantage of its usage is the removal of a destabilizing truncation error term that occurs in the original form of MAC.[1]

A finite-difference approximation to the vorticity is

$$\omega^n_{i+1/2,j+1/2} \equiv \frac{u^n_{i+1/2,j+1} - u^n_{i+1/2,j}}{\delta z}$$
$$- \frac{v^n_{i+1,j+1/2} - v^n_{i,j+1/2}}{\delta r}, \tag{3.14}$$

with centering at cell corners. Equations 3.11 and 3.12 can be combined to obtain a transport expression for $\omega_{i+1/2,j+1/2}$, which, like the differential equation, is independent of the θ field. Accordingly, the explicit calculation of the tilde velocities ensures that the vorticity at every internal mesh corner point is correct, independent of the choice of θ. This is not true, however, for corner points that lie on rigid walls, which are not correct until the tilde velocities have been corrected to ensure that D vanishes. For purely explicit calculations, which are acceptable for Reynolds numbers greater than about unity, SMAC vorticity diffusion from the wall is nevertheless correct, because the tilde velocities are based entirely on the final velocities from the previous cycle, which do agree with the proper wall vorticity. For an implicit formulation, the tilde velocity calculation would have to be combined with the potential function solution, the two being iterated simultaneously if a direct solution could not be obtained.

In the first of three phases for every calculational cycle, equations 3.11 and 3.12 are used to calculate $\tilde{u}$ and $\tilde{v}$ throughout the mesh. For the full cells,

$$\theta_{i,j} \equiv g_r r_i + g_z z_j \,,$$

Sec. 3B THE FINITE-DIFFERENCE EQUATIONS

where r_i and z_j are cell-centered coordinates in problem units. The choice of the θ field is based on solution efficiency and the mesh boundary conditions. No value for $\theta_{i,j}$ need be specified for cells outside the rigid walls; the normal velocity at each wall position is simply set equal to zero. In every surface cell, the value of $\theta_{i,j}$ is given by the expression in the original SMAC method.

$$\theta_{i,j} = \theta_{i,j} \text{ (applied)} + \frac{2\nu}{\delta z}(v_{i,j+1/2} - v_{i,j-1/2}) \,.$$

An improved treatment is possible with surface markers as described later. Tilde velocities are not, however, calculated for the empty cell faces of the surface cells; instead, they will be determined later in the cycle in such a way as to make $\tilde{D}_{i,j}$ vanish for each surface cell.

As a result of this first phase of the cycle, the vorticity has been correctly implanted into the new velocity field, which, however, is everywhere faulty in the values of $\tilde{D}_{i,j}$. Even the summed value of $\tilde{D}_{i,j}$ over the entire fluid region does not vanish unless the fluid is completely confined by rigid walls.

The function of the second phase is to convert the tentative (tilde) velocity field into the final velocity field for the cycle so that $D_{i,j} = 0$ for every cell. This must occur in such a way as to preserve identically the vorticity deposited into the field in the first phase. Accordingly, the change in every velocity must be given by the gradient of a potential function, ψ.

$$u^{n+1}_{i+1/2,j} = \tilde{u}^{n+1}_{i+1/2,j} - \frac{1}{\delta r}(\psi_{i+1,j} - \psi_{i,j}) \,, \text{ and}$$

$$v^{n+1}_{i,j+1/2} = \tilde{v}^{n+1}_{i,j+1/2} - \frac{1}{\delta z}(\psi_{i,j+1} - \psi_{i,j}) \,. \tag{3.15}$$

THE TWO-DIMENSIONAL NAVIER-STOKES MODEL

From equations 3.10 and 3.15, it follows that

$$D_{i,j}^{n+1} = \tilde{D}_{i,j} - \frac{1}{r_i^\alpha \delta r^2}\left[r_{i+1/2}^\alpha(\psi_{i+1,j} - \psi_{i,j})\right.$$

$$\left. - r_{i-1/2}^\alpha(\psi_{i,j} - \psi_{i-1,j})\right]$$

$$- \frac{1}{\delta z^2}(\psi_{i,j+1} + \psi_{i,j-1} - 2\psi_{i,j}), \qquad (3.16)$$

which, together with the requirement $D_{i,j}^{n+1} \equiv 0$ and appropriate boundary conditions, serves to determine uniquely the value of $\psi_{i,j}$ for every cell.

For the boundary condition near a rigid wall,

$$\psi_{i-1,j} = \psi_{i,j},$$

which ensures that the normal velocity at the wall will still vanish after the transformation. For free surface cells, we take $\psi_{i,j} = 0$. For a continuative outflow wall, the outside cell has $\psi_{i-1,j} = 0$. In this case, the value of the normal velocity on the wall is not calculated in the tilde-velocity phase, but is chosen instead to equal the tilde velocity on the opposite face of the interior cell. After using this to calculate $\tilde{D}$ and the ψ values, the final wall velocity for the new cycle is calculated by means of equation 3.15.

The equation for ψ can be solved by direct methods, but it often will be necessary to use an iteration procedure. Substitution of the ψ solution into equation 3.15 then results in the final set of velocities for every position, except the empty cell faces of the surface cells, which now are chosen in such a way as to make $D_{i,j}^{n+1} = 0$ for each of those cells.

Sec. 3B   THE FINITE-DIFFERENCE EQUATIONS

This completes the essential parts of the calculational cycle. In the original MAC and SMAC versions, all pressures or pseudopressures were specified at cell centers. Chan and Street[3] developed a technique for more accurate delineation of the free surface. This permitted the pressure equation to be adjusted so that the free surface pressure could be specified at the surface itself, rather than at the center of the surface cell. Nichols and Hirt[4] modified the Chan procedure and devised a technique for defining the fluid surface by a set of surface marker particles that move with the local fluid velocity. These particles allow surface cell pressures in MAC to be accurately specified by means of linear interpolation (or extrapolation) between the known values of pressure in the nearest full cell and the desired fluid surface pressure. The Nichols and Hirt scheme uses a set of special marker particles spaced at intervals along the surface, the exact surface configuration being given by line segments joining the particles in sequential order. These surface markers are moved with the local fluid velocity. The value of some field variable at the center of a surface (SUR) cell, is based on a linear interpolation or extrapolation using the known values at the free surface and the nearest full (FUL) cell center. The interpolated value is a function of the distance from the actual surface to the center of the nearest FUL cell, along a line connecting the centers of the SUR and FUL cells. This distance is denoted by d in Figure 3.2. The general form of the interpolation is

$$q_{\text{SUR}} = (1 - \eta) q_{\text{FUL}} + \eta q_{\text{free surface}},$$

where q is some cell-centered variable and η is the ratio of the distance between cell centers to the distance d. When the SUR cell has more than one FUL neighbor, as in Figure 3.2, the interpolation process chooses the neighbor whose intersection point lies nearest the center of the SUR cell.

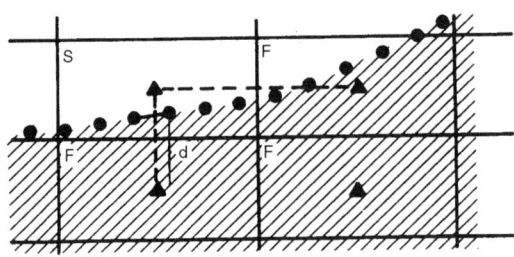

Fig. 3.2. The interpolation neighbor of the SUR cell is the FUL cell below it.

3C. Application to Tsunami Wave Propagation

The application of the two-dimensional incompressible Navier-Stokes model to the problem of tsunami waves interacting with continental slopes and shelves was first reported by Mader.[9,10]

Present evidence suggests that tsunami waves consist of a train of several large, approximately sinusoidal waves of about 1 meter in height moving in the deep ocean at approximately the shallow water speed of $(gD)^{1/2}$ or 210 meters sec^{-1} at the average Pacific Ocean depth of 4500 meters, having periods of 10–30 minutes, wave lengths of 200–600 kilometers, and numerous smaller waves. It is the first four or five large waves that are of primary interest.

The objectives of the study were to determine whether time dependent, nonlinear, viscous calculations of an incompressible fluid could be performed for gravity waves with the extreme height-to-width ratios ($\sim 1/100{,}000$) of tsunami waves and to determine if the growth of the waves could be followed as they interact with the continental slope. Another objective of the study was to compare the results of the Navier-Stokes model with shallow water, long wave calculations for identical problems.

ZUNI calculations were performed for waves that resemble tsunami waves. Street et al.[8] used the MAC technique to numerically simulate long water waves and found that they could numerically reproduce the observed propagation of solitary waves in a horizontal

Sec. 3C APPLICATION TO TSUNAMI WAVE PROPAGATION

channel and the runup of a solitary wave on a vertical wall. The calculations of Street et al. did not consider waves of the height-to-width ratios of tsunami waves. Similar conditions were performed using the *ZUNI* code, and the vertical wall runup results agreed with the experimental and numerical results of Street et al.

Most of the calculations were performed with 15 cells in the Y direction and 68 cells in the X direction. The cells were rectangles 450 meters high ($\triangle Y$) the in Y direction and 6750 meters long ($\triangle X$) in the X direction. The cell aspect ratio was 1/15. The time increment used was 3 sec. The convergence criterion used was 0.02. The water level was placed at 4450 meters or 50 meters up into the eleventh cell. The gravity constant was -9.8 meters sec^{-2}. The viscosity coefficient used was 2.0 g sec^{-1} meters^{-1} (0.02 poise), a value representative of the actual viscosity for water. Values between 200 and 2.0 were tried, and the viscosity did not significantly affect the results. This is expected because the actual energy dissipation owing to viscosity is only 1 in 10^7, for a viscosity of 200.

The stability requirements for the SMAC type of calculation are discussed in reference 1. Because the wave front must not pass through more than one cell in one time step, we have the stability criterion

$$C \triangle t < 2(\triangle X)(\triangle Y)/(\triangle X + \triangle Y),$$

where C is wave speed, $\triangle X$ and $\triangle Y$ are cell widths, and $\triangle t$ is the time increment. For the tsunami calculation described previously, $\triangle t$ must be less than 4 sec. We ran with 3 sec and had no evidence of instability; however, attempts to run with a time step of 9 sec always resulted in unstable numerical results that quickly turned to nonsense. Although Street et al.[8] report that the MAC numerical method was observed to be stable for zero viscosity, analysis by Nichols and Hirt[4] suggests the perturbations could grow if the viscosity was not larger than about 20 and smaller than 30,000 g sec^{-1} m^{-1}. We could determine no difference between a viscosity of 200 and 2.0 for the tsunami calculation; so if perturbations are growing, they must be doing so at a rate smaller than the errors associated with the iteration convergence criterion.

109

THE TWO-DIMENSIONAL NAVIER-STOKES   Chap. 3
MODEL

The original *ZUNI* code was written for the CDC 6600 and the 7600 computers. Because of the smaller amount of significance available on most computers, it was advantageous to use the actual hydrostatic pressure for the full cells rather than pseudopressures (generally zero), characteristic of the SMAC method to obtain adequate convergence of the iteration process. The convergence criterion for the SMAC iteration was defined as the maximum permitted change in pressure from hydrostatic pressure in any cell between iteration steps divided by the sum of the changes at the two iteration steps.

The *ZUNI* code includes features for prescribing the particle velocity along the left boundary as a sinusoidal function of time:

$$U = A \sin\Big(B(\text{TIME})\Big) \quad and \quad U = A\Big[\sin\Big(B(\text{TIME})\Big)\Big]^2,$$

where A for shallow water Airy waves is HC/D, where H is wave height, C is wave speed, and D is depth; and B is $2\pi/T$, where T is the period.

The dimensions of the shoaling calculations with a continental slope of 1:15 ending in a shelf 500 meters deep are sketched in Figure 3.3.

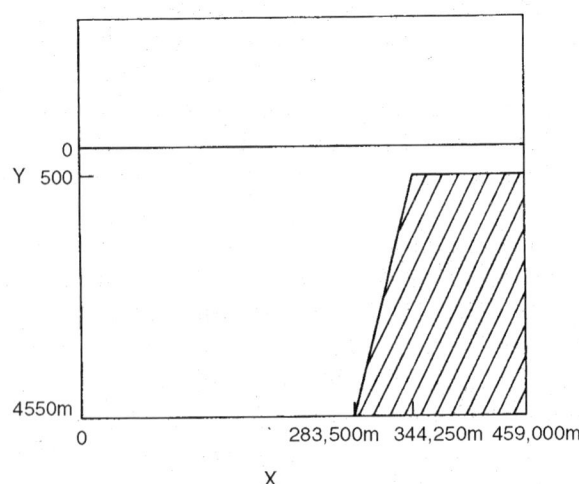

Fig. 3.3. A sketch of the geometry of the calculation.

Sec. 3C   APPLICATION TO TSUNAMI WAVE
                                              PROPAGATION

Solitary Tsunamis

Although single solitary like waves are not realistic models of tsunami waves, they are useful for demonstrating fundamental features of the flow and for checking numerical results.

The waves were generated by prescribing as a function of time the velocity of water flowing in from the left boundary. The velocity in the X direction was prescribed as

$$U = 0.04666 \left[\sin\Big(0.004713(\text{TIME})\Big)\right]^2,$$

where TIME is in sec. In 4550 meters deep water, this results in a single wave above the surface with a height of ~ 1 meter, a width of ~ 140 kilometers, a shallow water speed of 210 meters $\sec^{-1}$, and a period of ~ 660 sec.

The computed wave surface profiles for the 140 kilometers wide single wave interacting with a 1:15 continental slope, running along a 500 meters deep continental shelf and reflecting off a cliff are shown in Figure 3.4. In Figure 3.5, the profiles are shown for multiple waves.

As a wave proceeds up the continental slope, the wave period remains constant while the velocity decreases from approximately 210 to 70 meters $\sec^{-1}$, the wave length decreases from approximately 140 to 47 kilometers, the height increases from 0.95 to 1.6 meters, and then slowly decreases forming a complicated train of waves. Upon reflection from the right boundary, a wave 2.4 meters high is formed, which is approximately double the height of the wave before it arrived at the boundary.

The experimental and numerical results for the solitary, vertical wall runup waves of Street et al.[8] show that at small values of wave height divided by depth, the slope of the wave runup vs wave height is almost 2.0. The computed wave surface profiles for the 140 kilometers wide single wave running up a 1:15 continental shelf to above still water level are shown in Figure 3.6. Figure 3.7 shows the maximum shoaling amplitude of the wave as a function of shoaling depth. Also shown is the shallow water curve for the same example.

THE TWO-DIMENSIONAL NAVIER-STOKES MODEL   Chap. 3

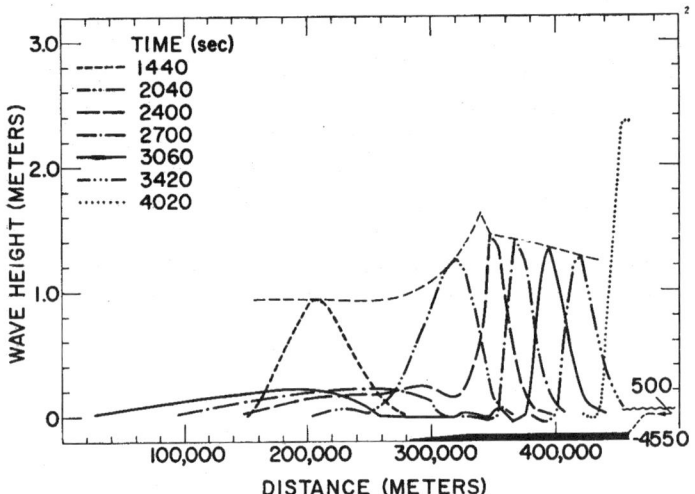

Fig. 3.4. Computed wave surface profiles for the 140 kilometers wide single wave interacting with a 1:15 continental slope (sketched on bottom of graph), a continental shelf 500 meters deep and reflecting off a cliff.

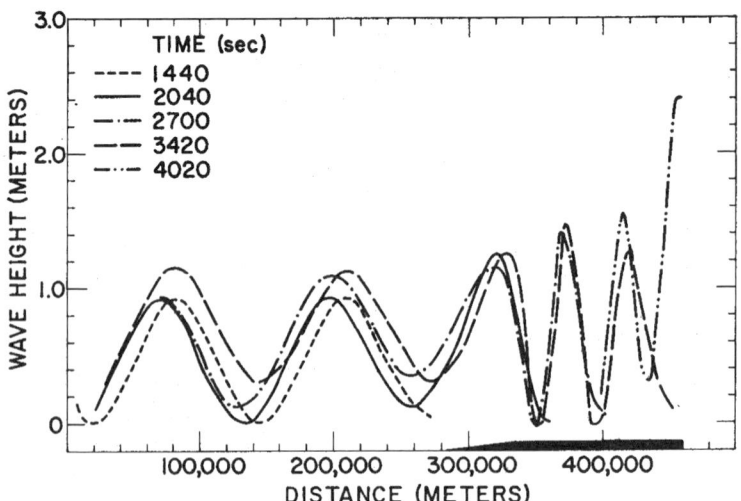

Fig. 3.5. Computed wave surface profiles for multiple waves at various times for the system in Figure 3.4.

Sec. 3C	APPLICATION TO TSUNAMI WAVE PROPAGATION

The growth of the wave was compared with the experimental and theoretical results of Madsen and Mei[11] for solitary waves interacting with uneven bottoms. Peregrine[12] treated a similar problem, but with less appropriate initial conditions. The results of Madsen and Mei shoal only to depths of 0.2 of the initial depth. Within this range, their results are in reasonable agreement with the results shown in Figures 3.4 and 3.5. The maximum height of the wave was 2.8 meters or about three times the initial wave height of 0.95 meter. This must be considered a lower limit, because the resolution of the calculation is inadequate to determine the maximum height.

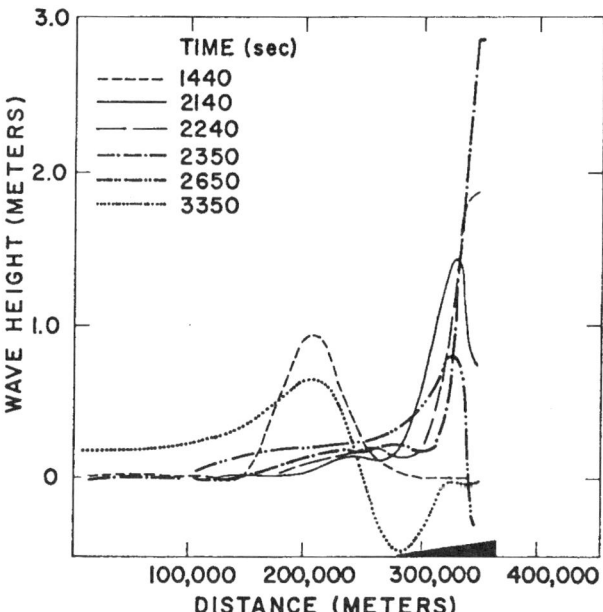

Fig. 3.6. Computed wave surface profiles for a single wave interacting with a 1:15 continental slope.

One approach to this problem is to look only at 100 meters of water and compress the scale of the calculation by 45. As input, we used a piston that would initially produce a wave of the height calculated in the previous problem at 100 meters. As shown in Figure 3.6, the wave height is about 1.8 meters. The wave speed is closely approximated by $(gD)^{1/2}$ or 31.5 meters sec^{-1}.

113

THE TWO-DIMENSIONAL NAVIER-STOKES MODEL   Chap. 3

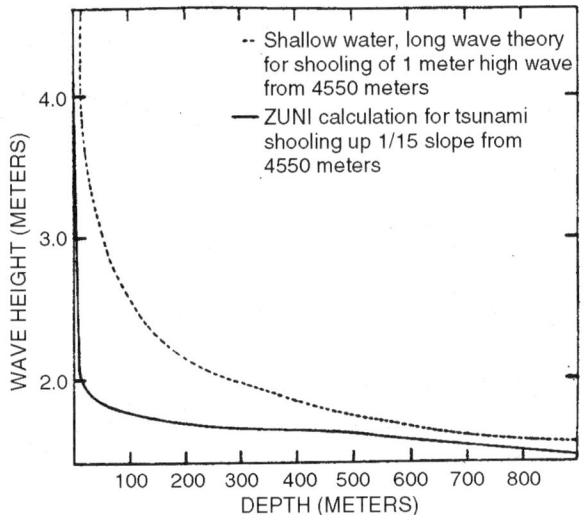

Fig. 3.7. The amplitude of calculated 1 meter half-height, 1320 sec (or 1 meter high, 660 sec single wave above surface) tsunami waves as they shoal up a 1:15 slope from 4550 meters. Also shown is the shallow water, long wave curve.

For a single wave above the surface, the period is 660 sec and the wave length 20.8 kilometers; for a sine wave, the period is 1320 sec and the wave length 41.6 kilometers.

The calculation was performed with a mesh of 15 x 68 cells, as before, with a $\triangle X$ of 150, $\triangle Y$ of 10, $\triangle t$ of 0.5, and a convergence error of 0.002; the 1:15 slope was started at the 42nd cell or at 6300 meters. The depth was 101.1 meters and

$U = 0.567 \sin[0.004713(\text{TIME})]$.

The wave profiles are shown in Figure 3.7, and the maximum wave height is 3.47 meters, or about 1.9 times the 1.8 meters wave height used initially in the calculation and 3.6 times the 0.95 meter wave height before any shoaling had occurred. The wave then flattens to a height of 2.75 ± 0.1 meters over the width considered. This is close to the 2.8 meters value calculated by the less resolved calculation. This also explains the flat top observed upon reflection in the less resolved calculation.

Sec. 3C   APPLICATION TO TSUNAMI WAVE
                                             PROPAGATION

It is apparently a realistic average height for the cell size used, and the flat top over several cell widths is also apparently correct.

For a more realistic model of a tsunami, we used sine waves with a half-height of 0.5 or 1 meter, a period of 1320 or 660 sec, and a wave length of 280 or 140 kilometers.

One-Meter Half-Height, 1320 sec Tsunamis

Using a piston velocity prescribed by

$$U = 0.04666 \sin[0.004713(\text{TIME})] \, ,$$

a wave develops in 4550 meters deep water of approximately 0.95 meter half-height, with a wave length of ~ 280 kilometers, a period of 1320 sec, and a speed of 210 meters sec^{-1}. The wave is slightly dispersive as it proceeds up a constant depth channel.

The wave surface profiles were computed for the tsunami interacting with a 1:15 continental slope, running along a 500 meters deep continental shelf and reflecting off a cliff. The profiles are similar to those shown in Figures 3.3 and 3.4. The peak shoaling height is 1.6 meters, which decreases to 1.4 meters before it reflects off the wall with maximum and minimum heights of +2.75, −3.4, and +2.75 meters.

The calculated wave surface profiles are shown in Figures 3.8–3.10, along with the shallow water, long wave calculations using the *SWAN* code for the identical model. The long wave results do not disperse in the deep channel, they do shoal higher and steeper, and they do not disperse as they run along the shallow channel. Experimental evidence that such waves should disperse in the shallow channel is given by Madsen and Mei.[11] The wave that reflects from the wall in the *SWAN* calculation is 20 percent higher than the reflected wave in the *ZUNI* calculation.

Figure 3.11 shows the computed wave surface profiles as the tsunami interacts with a 1:15 continental slope. The maximum and minimum calculated heights are +2.83, −4.0, +3.76, −4.11, and +4.04 meters, so the maximum runup does not result from the shoaling of the first wave, but from the shoaling of the second or third wave.

115

THE TWO-DIMENSIONAL NAVIER-STOKES MODEL

Chap. 3

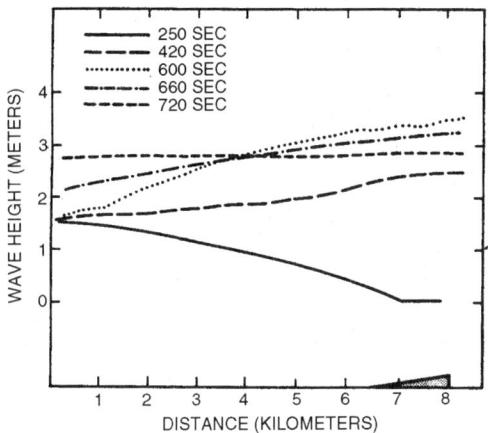

Fig. 3.8. Computed wave surface profiles for a 1.8 meters half-height, 1320 sec tsunami shoaling up a 1:15 slope from 101.1 meters.

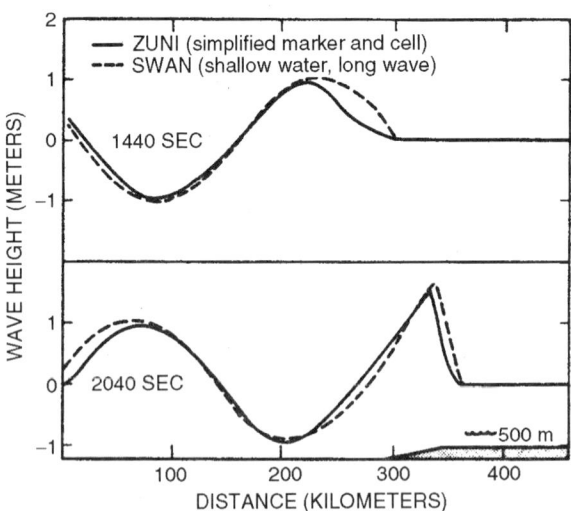

Fig. 3.9. Computed wave surface profiles for a 1 meter half-height, 1320 sec tsunami interacting with a 1:15 continental slope, a continental shelf 500 meters deep and reflecting off a cliff; and the shallow water, long wave calculations for the same model.

Sec. 3C APPLICATION TO TSUNAMI WAVE PROPAGATION

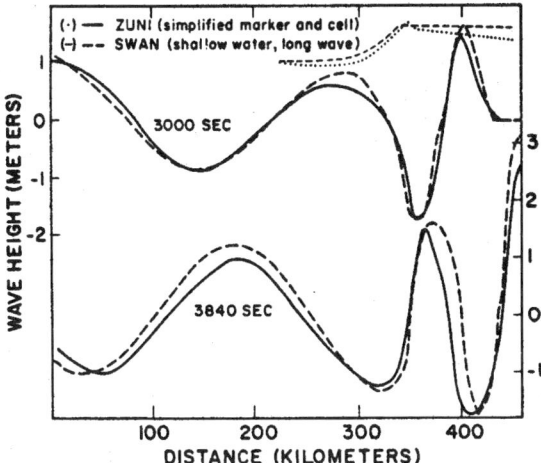

Fig. 3.10. Continuation of Figure 3.9.

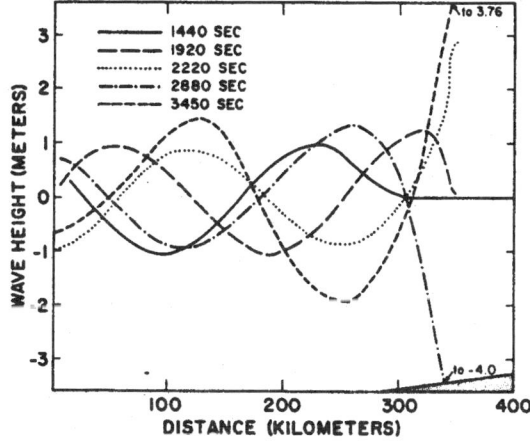

Fig. 3.11. Computed wave surface profiles for the wave in Figure 3.9 interacting with a 1:15 continental slope.

Such behavior has been observed for real tsunamis. We can now place an upper limit of tsunami wave growth from shoaling up a 1:15 continental slope of at least 4.0. As discussed earlier and shown in Figure 3.8, the scale of the calculation is such that this should be considered an average over the cell size used rather than the actual maximum values.

THE TWO-DIMENSIONAL NAVIER-STOKES   Chap. 3
MODEL

For a tsunami interacting with a 1:15 continental slope, running along a 950 meters deep continental shelf and reflecting off a cliff, the peak shoaling height is 1.5 meters, which decreases to 1.32 meters before it reflects off the wall with maximum and mimimum heights of $+2.56$, -3.0, and $+2.48$ meters.

One-Meter Half-Height, 660 sec Tsunamis

Using a piston velocity prescribed by

$$U = 0.04666\sin[0.009426(\text{TIME})] \, ,$$

a wave develops in 4550 meters deep water of approximately 0.86 meter half-height. The shorter wave length wave is more dispersive than the longer one discussed previously.

The wave surface profiles were computed for the tsunami interacting with a 1:15 continental slope, running along a 500 meters deep continental shelf and reflecting off a cliff. The peak shoaling height is 1.4 meters, which decreases to 0.68 meter before it reflects off the wall to a first-wave maximum of 1.15 meters. Subsequent wave interactions result in much larger wave runups.

The wave surface profiles were computed for the tsunami interacting with a 1:15 continental slope. The maximum calculated heights are $+2.14$, -3.56, $+3.44$, -3.9, and $+3.6$ meters. Again, the upper limit of tsunami wave growth from shoaling is at least a factor of 4.0.

The surface wave profiles were calculated for a 1.8 meters half-height, 660 sec tsunami shoaling up a 1:15 slope from 101.1 meters. Comparison with the wave profiles shown in Figure 3.8 shows that, with the same height at 101 meters, the smaller wave length shoals to a higher level but for a shorter distance and time.

Half-Meter Half-Height, 600 sec Tsunamis

Using a piston velocity prescribed by

Sec. 3C   APPLICATION TO TSUNAMI WAVE PROPAGATION

$$U = 0.02333 \sin[0.009426(\text{TIME})],$$

a wave develops in 4550 meters deep water of about 0.4 meter half-height.

The wave surface profiles were computed for the tsunami interacting with a 1:15 continental slope, running along a 500 meters deep continental shelf and reflecting off a cliff. The peak shoaling height is 0.62 meter, which disperses as it runs along the shelf to 0.28 meter. The wave runup heights are $+0.4$, -1.32, $+1.10$, -2.10, $+1.10$, and -1.72 meters.

Some of the calculated wave surface profiles are shown in Figure 3.12, along with the shallow water, long wave calculations using the *SWAN* code for the identical model. The difference between the two calculations increases as the wave progresses, differing by a factor of 2 upon reflecting from the wall. Although it could be stated that this is an example where the long wave, shallow water assumptions lead to appreciable error, this is not necessarily true because we do not know the nature of tsunami waves well enough to determine if the tsunami should be more like the long wave model or the one we used.

Because most tsunami waves that have been observed after traveling across the ocean have periods longer than 10 minutes, it is tempting to postulate that this is because the shorter waves are so dispersive that they cannot propagate long distances.

The surface wave profiles were computed as the tsunami interacts with a 1:15 continental slope. The maximum and minimum calculated are $+0.91$, -2.03, $+1.5$, -1.79, $+1.68$, -2.1, and $+1.54$ meters. Again, the upper limit of the tsunami wave growth from shoaling is at least a factor of 4.0.

Conclusions

The detailed numerical simulation of gravity waves that resemble the profile of actual tsunami waves was achieved. The interaction of tsunami waves with slopes that resemble the continental slope was realistically simulated. Wave heights were observed to increase by a factor of 4 as they shoaled up a 1:15 continental slope. The second or third wave often exhibited the highest wave runup.

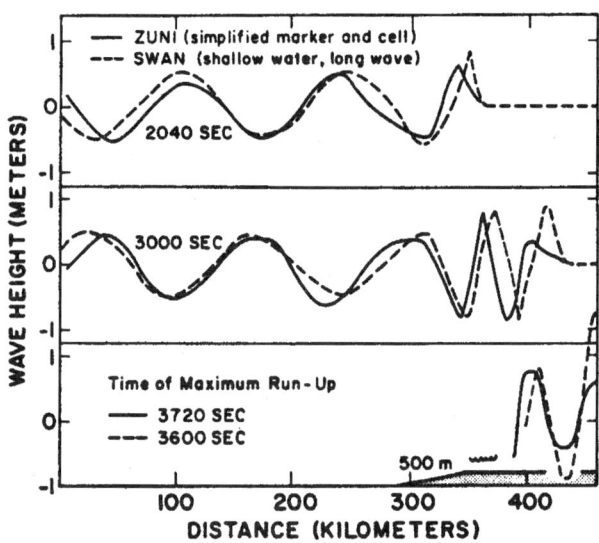

Fig. 3.12. Computed wave surface profiles for a 0.5 meter half-height 660 sec tsunami interacting with a 1:15 continental slope, a continental shelf 500 meters deep and reflecting off a cliff; and the shallow water, long wave calculations for the same model.

Similar results can be obtained using shallow water, long wave theory for long wave length tsunamis, but the theory is inadequate for short wave length tsunamis. The long wave tsunamis do not disperse in the deep channel, they shoal higher and steeper, and they do not disperse as they run along the shallow channel.

3D. Application to Underwater Barriers

A submerged barrier usually absorbs some of the wave energy by causing the wave to break prematurely and by reflecting part of the wave energy back seaward. Tsunami waves are of sufficiently long wavelength that they do not break, so underwater barriers will be effective only as reflectors of the energy. The shallow water, long wave theory is inadequate to describe the effect of underwater barriers on tsunami waves because the vertical component of velocity is a crucial feature of the flow.

Sec. 3D APPLICATION TO UNDERWATER BARRIERS

Johnson et al.[13] present results of an experimental investigation of the damping action of submerged rectangular breakwaters. They use the inshore wave height divided by the seaward wave height before interaction with the barrier as the transmission coefficient, and they graph it against the dimensionless quantity of barrier height divided by channel water depth.

The calculations were performed assuming the barrier was located in 101.1 meters of water and that it extended to within 21.1, 11.1, and 6.1 meters of still water surface. The wave height of 1.8 meters at 100 meters from Figure 3.7 was assumed. The wave period assumed was 1320, 660, and, for a more detailed wave interaction calculation, a short period of 110 sec.

The 21.1 and 11.1 meters deep barriers used the mesh described previously for Figure 3.8. The barrier was 450 meters wide, occupying cells 30–32 or from 4350–4800 meters. The position of the barrier is shown on the bottom of the figures.

The 6.1 meters deep barrier calculation was performed using a mesh of 23 by 45 cells with a $\triangle X$ of 150, $\triangle Y$ of 5.0, $\triangle t$ of 0.3, and a convergence error of 0.002; the barrier was 450 meters wide, occupying cells 20–22 or from 2850–3300 meters.

The surface wave profiles for a 660 sec period tsunami interacting with barriers 21.1, 6.1, and 11.1 meters below the water surface are shown in Figures 3.13–3.15. The initial inshore wave heights are 1.4, 1.1, and 0.54 meters, for the 21.1, 11.1, and 6.1 meters deep barriers, respectively, for an undisturbed seaward height of 1.84 meters. The 6.1 meters height is so far below the seaward height that it is probably a lower limit value.

Figure 3.15 also compares the surface wave profile for a barrier 11.1 meters below the water surface with shallow water, long wave calculations. As expected, the long wave model is inadequate.

The surface wave profiles were calculated for a 1320 sec period tsunami interacting with 11.1 meters deep barrier. The initial inshore wave height is 1.03 meters. This is probably a lower limit, because the seaward height is significantly larger when the calculation was ended because it was being disturbed by the right boundary.

The surface wave profiles were calculated for a 100 sec period tsunami interacting with barriers 21.1, 11.1, and 6.1 meters, below the water surface.

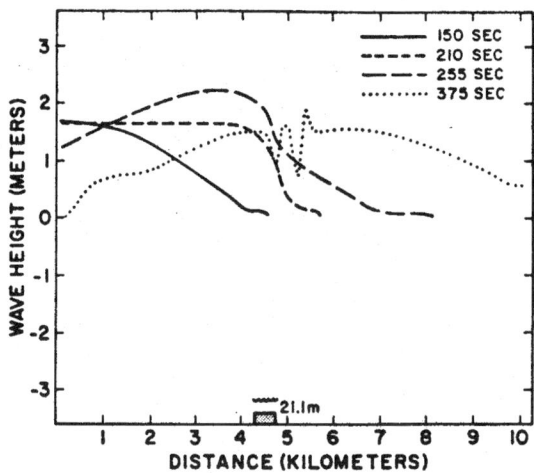

Fig. 3.13. Surface wave profiles for a 660 sec tsunami interacting with a 21.1 meters deep barrier.

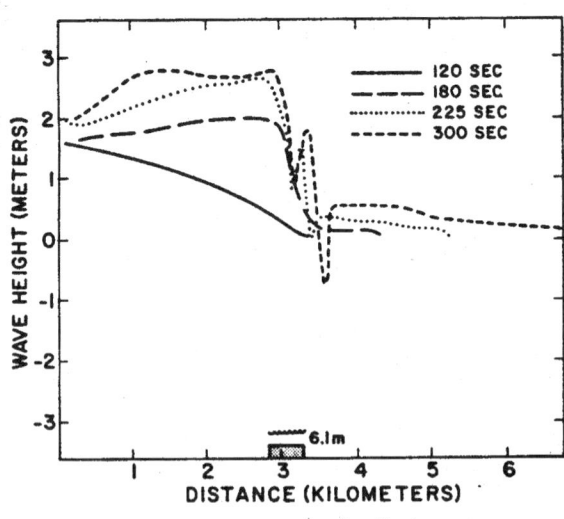

Fig. 3.14. Surface wave profiles for a 600 sec tsunami interacting with a 6.1 meters deep barrier.

Sec. 3D   APPLICATION TO UNDERWATER BARRIERS

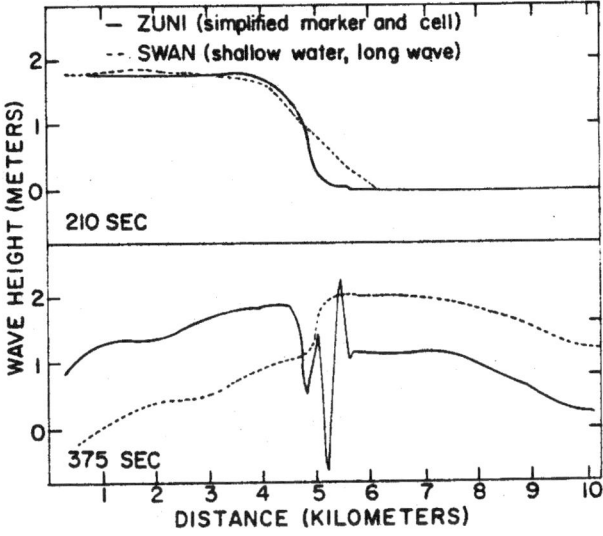

Fig. 3.15. Surface wave profiles for a 600 sec tsunami interacting with a 11.1 meters deep barrier, and the shallow water, long wave profiles for the same model.

The initial inshore wave heights are 1.34, 1.046, and 0.35 meters, respectively, for an undisturbed seaward height of 1.9 meters.

The transmission coefficient as a function of the barrier height divided by depth is shown in Figure 3.16. The experimental data from Figures 4 and 5 of Johnson et al.[13] and the tsunami curve of Figure 3.16 are shown in Figure 3.17. Although the characteristics are quite different between the experimental and calculated waves, the effectiveness of the underwater barrier appears similar.

Underwater barriers can reflect significant amounts of the tsunami energy. The shallow water, long wave theory is inadequate to describe the flow of tsunamis over underwater barriers.

THE TWO-DIMENSIONAL NAVIER-STOKES MODEL

Chap. 3

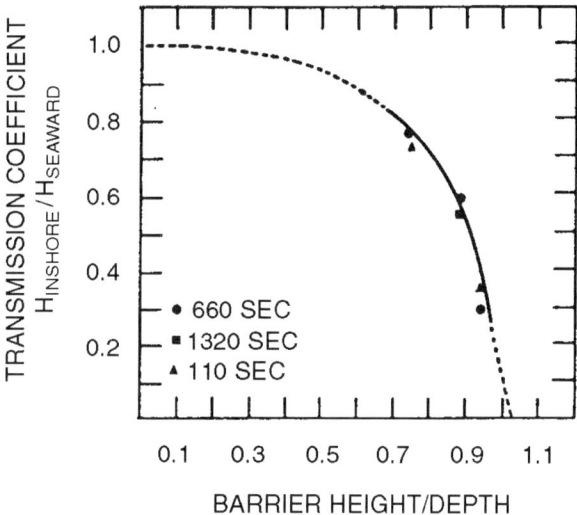

Fig. 3.16. The calculated transmission coefficient of tsunami waves as a function of barrier height divided by depth.

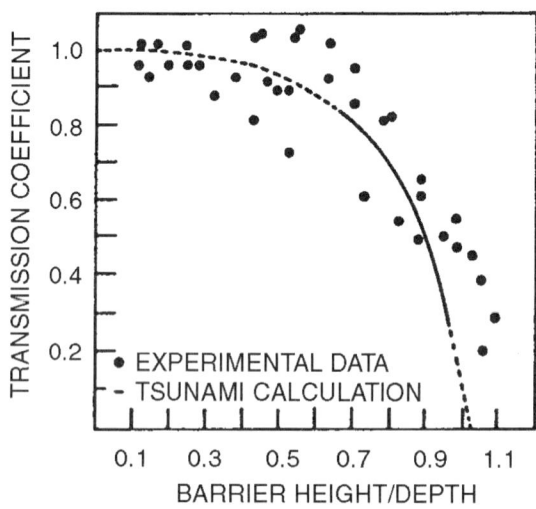

Fig. 3.17. The Johnson et al. experimental submerged breakwater data and the calculated tsunami curve in Figure 3.16.

3E. Waves from Cavities

The prediction of water waves generated by large-yield explosions has been based on extrapolation of empirical correlations of small-yield experimental data, usually assuming the waves were shallow water waves. Because the accuracy of such predictions is questionable, the need exists for a detailed description of the mechanism by which waves are generated by explosions. In particular, the "upper critical depth" phenomenon needs to be understood. The upper critical depth phenomenon is an experimentally observed wave height maximum that occurs when an explosive charge is approximately two-thirds submerged. The observed height at the upper critical depth is twice that observed for completely submerged explosive charges as shown in Figure 6.17 in Chapter 6. If the waves formed are shallow water waves capable of forming tsunamis, then the upper critical depth phenomenon is important to evaluating the probability of a tsunami event from other than tectonic events.

The early interaction of the detonation products of an explosive charge with the water and air interfaces and the resulting wave profile near the detonation has been theoretically evaluated using the multicomponent reactive compressible hydrodynamic code 2DE[7]. The late time flow could not be modeled using the 2DE code. In Chapter 6 the compressible flow for all the time of the experiments will be modeled using the *NOBEL* code.

The bubble is observed to increase to a maximum radius of about 0.5 meter in 0.2 sec and then take about 0.3 sec to collapse, as described in references 14 and 15, for a 1.27 centimeters radius PBX-9404 explosive sphere initiated at the center and immersed to a depth of 1.59 centimeters.

In reference 16, the results are described of an investigation of the flow after the cavity reaches its maximum dimensions of 0.5 meter in radius, both with and without a lip, using the shallow water, long wave code *SWAN* and the incompressible Navier-Stokes code *ZUNI*. The flow was assumed to be essentially incompressible at the times of interest, and the surrounding fluid was assumed to be approximately at rest at maximum bubble radius. This was a VERY incorrect assumption, as is shown using compressible flow modeling in Chapter 6.

THE TWO-DIMENSIONAL NAVIER-STOKES MODEL
Chap. 3

The wave observed by Craig[15] from the collapse of the bubble resulted in a train of waves moving at about 2.50 meters/sec with a 3.75 meters wavelength. Mass markers located 1 meter below the water surface and markers located 0.5 meter below the surface and 1 meter from the explosive showed no appreciable movement compared with those located nearer the surface or explosive charge. This result suggests that the fluid flow will not be well described by the usual shallow water, long wave model. The *SWAN* shallow water, long wave calculations were performed with 69 cells in the X direction and 130 cells in the Y direction. The cell sides were 0.06 meter and the time step was 0.001 sec. The gravity constant was -9.8 meters sec^{-1}.

The computed wave profiles using the shallow water equations in the *SWAN* code for the collapse of a 0.5 meter radius hole are shown in Figure 3.18. The wave profiles for the collapse of a 0.5 meter hole with a 0.25 meter high and 0.50 meter wide triangular lip (which approximates the experimentally observed bubble profile) are shown in Figures 3.19 and 3.20. Figure 3.20 also shows the velocity in the Y direction. The initial water depth was 3 meters.

The *ZUNI* incompressible Navier-Stokes calculations were performed with 100 cells in the X direction and 60 cells in the Y direction. The cell sides were 0.05 meter and the time step increment was 0.0003 sec. The convergence criterion is the maximum permitted change in pressure from hydrostatic pressure in any cell between iteration steps, divided by the sum of the changes at two iteration steps, and it was 0.01. Three or more surface particles were used in each cell. The gravity constant was -9.8 meters sec^{-2}. The viscosity coefficient was 0.01 gram sec^{-1}meter^{-1}. Preliminary calculations showed that the results were independent of whether the water depth was 3.0 meters or 1.5 meters; therefore, the cavity was chosen to be in water initially 1.5 meters deep. The wave amplitude at late times and the details of the bubble collapse after first collapse and jetting were sensitive to the amount of viscosity used in the calculation, so the value chosen was the smallest that would also permit numerically stable results. The computed wave profiles using the *ZUNI* code to solve the incompressible Navier-Stokes equations for the collapse of a 0.5 meter radius hole with a 0.25 meter high and 0.5 meter wide triangular lip are shown in Figure 3.21.

Sec. 3E   WAVES FROM CAVITIES

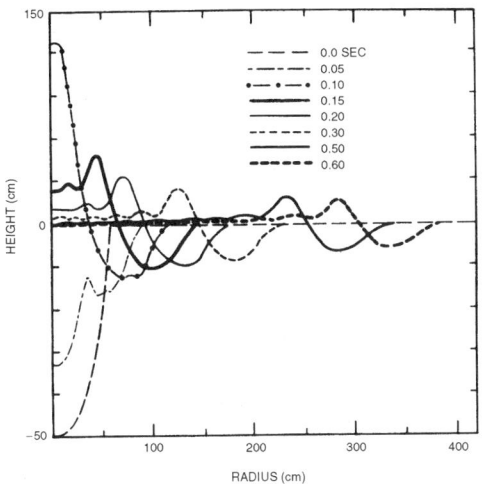

Fig. 3.18. The computed surface height vs radius at various times of the collapse of a 50 centimeters radius hole in 3 meters of water using the shallow water, long wave model and the *SWAN* code.

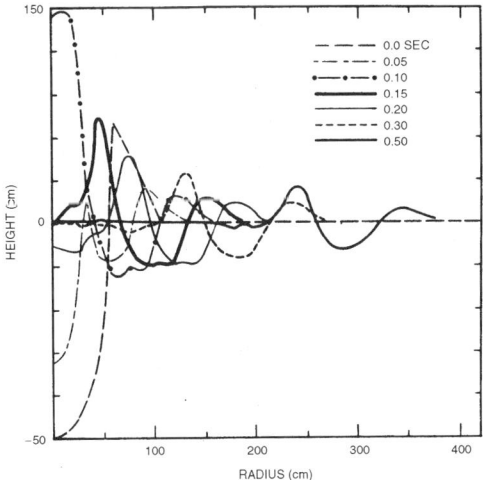

Fig. 3.19. The calculated surface height vs radius at various times of the collapse of a 50 centimeters radius hole with a triangular lip using the shallow water *SWAN* code.

THE TWO-DIMENSIONAL NAVIER-STOKES MODEL Chap. 3

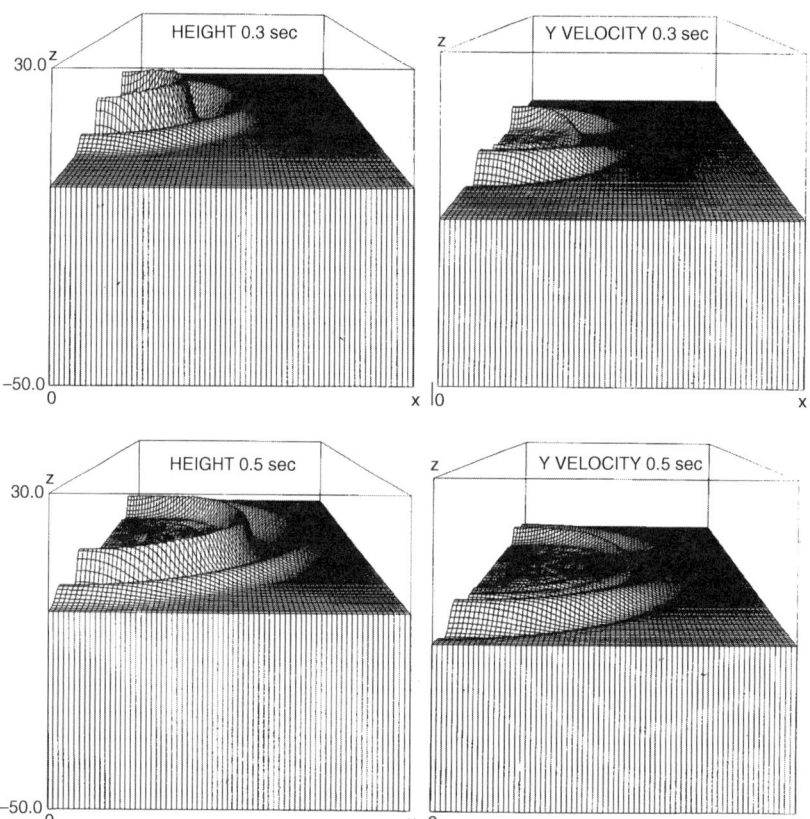

Fig. 3.20. Picture plots in three dimensions of the surface profiles, and the velocity in the Y direction profiles at 0.3 and 0.5 sec for the calculation described in Figure 3.19.

The collapse of the cavity is quite different between the *SWAN* and *ZUNI* calculation, as shown in Figures 3.19 and 3.21. The shallow water cavity calculation collapses from the side in less than 0.1 sec while the Navier-Stokes cavity calculation collapses from the bottom in about 0.5 sec. The experimentally observed bubble collapses approximately symmetrically from the bottom in about 0.3 sec, so the Navier-Stokes calculation is a more realistic description of the observed flow.

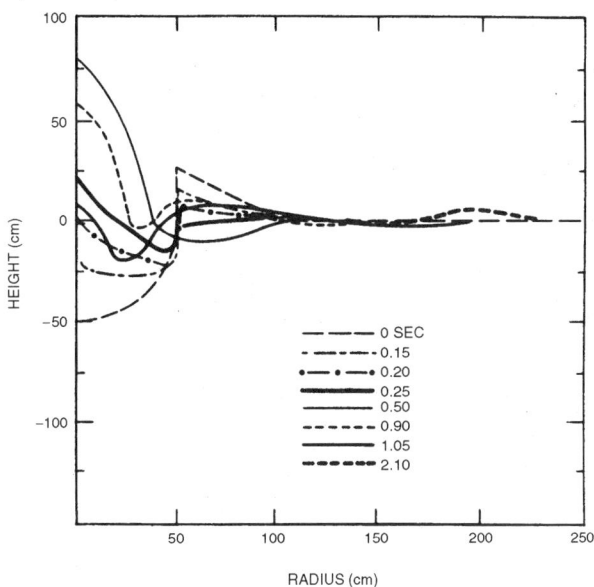

Fig. 3.21. The calculated surface height vs radius at various times of the collapse of a 50 centimeters radius hole with a triangular lip using the incompressible Navier-Stokes model and the *ZUNI* code.

The Navier-Stokes calculation results in wave parameters more closely approximating those observed experimentally than those calculated using the shallow water theory. The waves calculated using the incompressible Navier-Stokes equations have complicated wave patterns and the waves quickly decay into deep water waves, with the particle motion rapidly decreasing with increasing depth below the water surface. This is also in agreement with the mass marker observations.

The experimentally observed waves from the cavities formed by explosions near the water surface are better reproduced by models solving the incompressible Navier-Stokes equations than by models solving the shallow water, long wave equations. The experimentally observed waves are deep water waves. The observed upper critical depth phenomenon is apparently a result of a partition of energy near the water surface, which results in high-amplitude, deep water waves (of high potential and low kinetic energy) and not the shallow water waves required for tsunamis.

When the explosive is detonated under the water surface, more of the energy is imparted to the water resulting in waves that have smaller amplitude, but more of the energy is present in the water as kinetic energy rather than potential energy. The waves formed will more closely resemble shallow water waves and will not disperse as rapidly as waves formed from a surface detonation.

The upper critical depth phenomenon is apparently not important to the formation of tsunami waves. Large cavities located under the surface of the ocean will be more likely to result in shallow water tsunami waves than cavities on the surface. A cavity would have to be quite large and reach very deep into the ocean before it could be an effective agent for forming significant tsunami waves. To obtain a more realistic solution to the problem of wave formation from explosions, one needs to follow the hydrodynamics from the early compressible to late incompressible stages. Of particular interest is to determine the source and significance for wave formation of the water jet and root experimentally[15] observed before the explosive bubble reaches maximum radius at depths similar to the upper critical depth.

This was finally accomplished in December of 2002 after 30 years of unsuccessfully trying to model the complicated physics using the numerical modeling techniques and reactive, compressible hydrodynamic codes as they were developed. The results are described in reference 17, Chapter 6 and a PowerPoint presentation is on the NMWW CD-ROM with experimental and computer movies in the directory /NOBEL/CAVITY.

The Loihi Cone Collapse

Loihi is a volcanic undersea island just off the Island of Hawaii. It is growing toward the ocean surface and will become the next island in the Hawaiian chain.

The summit of Loihi collapsed during the Summer of 1996. The cavity generated by the collapse was 1000 meters wide and 300 meters deep in 1050 meters of water.

The initial wave length of a wave generated by a cavity 1000 meters wide is about the same as the width of the cavity, thus the ratio of the wave length to depth was about one.

Sec. 3E WAVES FROM CAVITIES

Since the depth of the ocean increases with increasing distance from the center of Loihi, the wave length-to-depth ratio becomes less than half as the wave travels from its source. Such a wave is a deep water wave (defined as wave length/depth less than 0.5) rather than a tsunami or shallow water wave (defined as wave length/depth greater than 20).

If the shallow water model is used to model the wave generated by the collapse cavity, the calculated wave amplitude and velocity will be much too large and the wave period will be too small.

The generation and propagation of the wave formed by a 500 meters radius cavity, 300 meters deep in 1050 meters deep water was modeled using the *ZUNI* two-dimensional Navier-Stokes code.

The mesh used to model the wave generation and propagation was 64 cells each 150 meters wide in the radial direction, and 16 cells each 140 meters high. The surface was described by surface particles initially located about 45 meters apart. The surface resolution was maintained by increasing the number of surface particles as needed. The time step was 0.5 sec. The wave generation and propagation process was insensitive to the size of the cells or the time step.

The cavity collapse occured from the bottom and the side of the cavity. A jet was formed at the center of the cavity which exceeded in height the initial depth of the cavity. A dispersive wave was formed with an initial height of 60 meters and a velocity of about 50 meters/sec (shallow water velocity is 100 meters/sec). The wave decayed to less than 8 meters height after 100 sec or 5.5 kilometers of travel as shown in Figure 3.22. The wave length increased to 6 kilometers and the velocity to 85 meters/sec. Since the wave generated by the Loihi cavity collapse traveled in water deeper than one kilometer, the Loihi wave had a smaller height and more deep water character than the modeled wave. The movies of the *ZUNI* calculations of the Loihi cone collapse tsunami are on the NMWW CD-ROM in the directory /TSUNAMI.MVE/LOIHI.MVE.

The wave generated by the collapse of 1000 meters wide, 300 meters deep infinitely long trough of water was modeled using the Linear Gravity Wave (*LGW*) model. The *LGW* model was used to model the formation of the surface cavity from an ocean floor displacement 1000 meters wide and 300 meters deep in 1050 meters water. The resulting initial surface cavity depth was only 150 meters and the cavity depth decreased with increasing width.

THE TWO-DIMENSIONAL NAVIER-STOKES MODEL Chap. 3

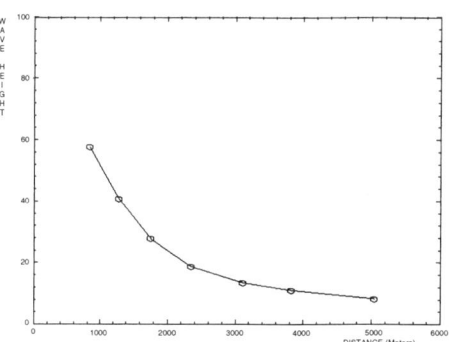

Fig. 3.22. The calculated surface wave height as a function of radial distance traveled for a *ZUNI* calculation of a 500 meters radius cavity, 300 meters deep in 1050 meters deep water.

Thus the initial surface cavity assumed for the *ZUNI* model, which simply reproduced the ocean floor displacement, was too large and the resulting wave heights and lengths are upper limits.

A full three-dimensional Navier-Stokes model such as *SOLA-3D* would be required to model the wave generated by the Loihi summit collapse accurately.

The Loihi cone collapse did not generate a tsunami wave. The wave was a deep water wave that rapidly dispersed as it propagated. It decayed as a deep water wave and became part of the normal ocean swell before it reached the coast of the island of Hawaii.

The Loihi volcano is often visited by the University of Hawaii submersible as part of their study of the process of formation of an island. They were surprised, in the Summer of 1996, to see that the cone had collapsed since their last visit. They wishfully concluded that it must have occurred slowly, since a tsunami wave was not detected on the Hilo tide gauge, and that they did not need to be concerned about such events being a hazard to the submersible. As shown in the above calculations, the collapse could have occurred rapidly and still not have generated a detectable tsunami wave. Such a rapid collapse would present a serious hazard to any submersible unfortunate enough to be present when it occurs.

References

1. J. E. Welch, F. H. Harlow, J. P. Shannon, and B. J. Daly, "The MAC Method: A Computing Technique for Solving Viscous, Incompressible, Transient Fluid Flow Problems Involving Free Surfaces," Los Alamos Scientific Laboratory LA–3425 (1965).
2. C. W. Hirt and J. P. Shannon, "Free Surface Stress Conditions for Incompressible Flow Calculations," Journal of Computational Physics, Vol. 2, pp. 403–411 (1968).
3. R. K. C. Chan and R. L. Street, "A Computer Study of Finite-Amplitude Water Waves," Journal of Computational Physics, Vol. 6, pp. 68–94 (1970).
4. B. D. Nichols and C. W. Hirt, "Improved Free Surface Boundary Conditions for Numerical Incompressible-Flow Calculations," Journal of Computational Physics, Vol. 8, pp. 434–448 (1971).
5. F. H. Harlow and A. A. Amsden, "A Simplified MAC Technique for Incompressible Fluid Flow Calculations," Journal of Computational Physics, Vol. 6, pp. 322–325 (1970) and Los Alamos Scientific Laboratory report LA–4370 (1970).
6. A. A. Amsden, "Numerical Calculations of Surface Waves: A Modified *ZUNI* Code with Surface Particles and Partial Cells," Los Alamos Scientific Laboratory report LA–5146 (1973).
7. Charles L. Mader, *Numerical Modeling of Detonations*, University of California Press, Berkeley, California (1979).
8. R. L. Street, R. K. C. Chan, and J. E. Fromm, "The Numerical Simulation of Long Water Waves on Two Fronts" in *Tsunamis in the Pacific Ocean*, W. M. Adams, Editor, University of Hawaii East–West Center Press, pp. 453–473, Honolulu, Hawaii (1969).
9. Charles L. Mader, "Numerical Simulation of Tsunamis," Hawaii Institute of Geophysics report HIG–73–3 (1973).
10. Charles L. Mader, "Numerical Simulation of Tsunamis," Journal of Physical Oceanography, Vol. 4, pp. 74–82 (1974).
11. O. S. Madsen and C. C. Mei, "Transformation of a Solitary Wave Over an Uneven Bottom," Journal of Fluid Mechanics, Vol. 39, pp. 781–791 (1969).
12. D. H. Peregrine, "Long Waves on a Beach," Journal of Fluid Mechanics, Vol. 27, pp. 815–827 (1967).

13. J. W. Johnson, R. A. Fuchs, and J. R. Morison, "The Damping Action of Submerged Breakwaters," Transactions of the American Geophysical Union, Vol. 32, pp. 704–718 (1951).
14. Charles L. Mader, "Detonations Near the Water Surface," Los Alamos Scientific Laboratory report LA–4958 (1972).
15. Bobby G. Craig, "Experimental Observations of Underwater Detonations near the Water Surface," Los Alamos Scientific Laboratory report LA–5548–MS (1974).
16. Charles L. Mader, "Calculations of Waves Formed from Cavities," Proceedings of 15th Coastal Engineering Conference, pp. 1079–1092 (1976).
17. Charles L. Mader and Michael L. Gittings, "Calculations of Water Cavity Generation," Science of Tsunami Hazards, Vol. 21, pp. 91–118 (2003).

4

THE THREE-DIMENSIONAL NAVIER-STOKES MODEL

The calculation of three-dimensional, nonlinear, incompressible viscous flow water waves may be performed using a modified marker and cell (MAC) finite-difference technique which uses pressure and velocity as the dependent variables. The surface height of the center of each cell is usually the extent of the resolution of the water surface available in three-dimensional codes.

The *SOLA-3* code is a three-dimensional version of the two-dimensional code described in references 1 and 2. The code is described in detail in reference 3 by Hirt, Nichols, and Stein. It requires a large computer and large scale graphic facilities. The resolution available is limited by the size of the computer memory so a variable mesh capability has been developed to improve the numerical resolution in the regions of interest. The surface height resolution is also limited to the height of the center of each cell which is computed from the equation

$$\frac{\partial H}{\partial t} + U_x \frac{\partial H}{\partial x} + U_y \frac{\partial H}{\partial y} = U_z \, .$$

The Navier-Stokes equations for incompressible viscous flow to be solved by finite difference are given in Chapter 1 as equations 1.4 and 1.7.

4A. The Finite-Difference Equations

The following description of the *SOLA-3D* numerical solution technique is taken from reference 3.

THE THREE-DIMENSIONAL NAVIER-STOKES MODEL Chap. 4

The differential equations to be solved are written in terms of Cartesian coordinates (x, y, z). For cylindrical coordinates (r, θ, z), the x coordinate is interpreted as the radial direction, the y coordinate is transformed to the azimuthal coordinate $r\theta$, and z is the axial coordinate. In addition, for cylindrical geometry several terms must be added to the Cartesian equations of motion. In the following, these terms are included with a coefficient α, such that $\alpha = 0$ corresponds to Cartesian geometry and $\alpha = 1$ corresponds to cylindrical geometry.

The mass continuity equation for an incompressible fluid is

$$\frac{\partial u}{\partial x} + R\frac{\partial v}{\partial y} + \frac{\partial w}{\partial z} + \alpha\frac{u}{x} = 0 . \tag{4.1}$$

The velocity components (u, v, w) are in the coordinate directions (x, y, z) or (r, θ, z), and the coefficient R depends on the choice of coordinate system in the following way. When cylindrical coordinates are used, y derivatives must be converted to azimuthal derivatives,

$$\frac{\partial}{\partial y} \longrightarrow \frac{1}{r}\frac{\partial}{\partial \theta} .$$

This transformation is accomplished by using the equivalent form

$$\frac{1}{r}\frac{\partial}{\partial \theta} = \frac{r_m}{r}\frac{\partial}{\partial y} , \tag{4.2}$$

Sec. 4A THE FINITE-DIFFERENCE EQUATIONS

where $y = r_m \theta$ and r_m is a fixed reference radius. In *SOLA-3D*, r_m is the maximum radius in the problem. Transformation 4.2 is particularly convenient because its implementation only requires a multiplier $R = r_m/r$ on each y-derivative in the original Cartesian coordinate equations. When Cartesian coordinates are to be recovered, R is set to unity and α is set to zero.

The equations of motion for the fluid velocity components in the three coordinate directions (u, v, w) are the Navier-Stokes equations:

$$\frac{\partial u}{\partial t} + u\frac{\partial u}{\partial x} + v\left(R\frac{\partial u}{\partial y}\right) + w\frac{\partial u}{\partial z} - \alpha\frac{v^2}{x}$$

$$= -\frac{\partial p}{\partial x} + g_x + f_x \;,$$

$$\frac{\partial v}{\partial t} + u\frac{\partial v}{\partial x} + v\left(R\frac{\partial v}{\partial y}\right) + w\frac{\partial v}{\partial z} + \alpha\frac{uv}{x}$$

$$= -\left(R\frac{\partial p}{\partial y}\right) + g_y + f_y \;, \text{ and}$$

$$\frac{\partial w}{\partial t} + u\frac{\partial w}{\partial x} + v\left(R\frac{\partial w}{\partial y}\right) + w\frac{\partial w}{\partial z}$$

$$= -\frac{\partial p}{\partial z} + g_z + f_z \;. \quad (4.3)$$

In these equations, p is the fluid pressure, fluid density has been taken as unity, (g_x, g_y, g_z) are body accelerations, and (f_x, f_y, f_z) are viscous accelerations. For a constant kinematic viscosity ν, the viscous accelerations are

THE THREE-DIMENSIONAL NAVIER-STOKES MODEL — Chap. 4

$$f_x = \nu \left[\frac{\partial^2 u}{\partial x^2} + R^2 \frac{\partial^2 u}{\partial y^2} + \frac{\partial^2 u}{\partial z^2} \right.$$
$$\left. + \alpha \left(\frac{1}{x} \frac{\partial u}{\partial x} - \frac{u}{x^2} - \frac{2R}{x} \frac{\partial v}{\partial y} \right) \right],$$

$$f_y = \nu \left[\frac{\partial^2 v}{\partial x^2} + R^2 \frac{\partial^2 v}{\partial y^2} + \frac{\partial^2 v}{\partial z^2} \right.$$
$$\left. + \alpha \left(\frac{1}{x} \frac{\partial v}{\partial x} - \frac{v}{x^2} + \frac{2R}{x} \frac{\partial u}{\partial y} \right) \right], \text{ and}$$

$$f_z = \nu \left[\frac{\partial^2 w}{\partial z^2} + R^2 \frac{\partial^2 w}{\partial y^2} + \frac{\partial^2 w}{\partial z^2} + \frac{\alpha}{x} \frac{\partial w}{\partial x} \right]. \quad (4.4)$$

Free surfaces are defined by their height $H(x, y, t)$ above the lower z boundary of the computing mesh. The time evolution of H is governed by the kinematic equation

$$\frac{\partial H}{\partial t} + u \frac{\partial H}{\partial x} + v \left(R \frac{\partial H}{\partial y} \right) = w. \quad (4.5)$$

The finite-difference mesh used for numerically solving the above equations consists of rectangular cells of width δx_i, depth δy_j, and height δz_k. The active mesh region consists of IBAR cells in the x direction, labeled with the index i, JBAR cells in the y direction labeled with the index j, and KBAR cells in the z direction labeled with the index k. This region is surrounded by a layer of fictitious or boundary cells used to get mesh boundary conditions. Thus, there are usually (IBAR + 2) (JBAR + 2) (KBAR + 2) total cells in a complete mesh. When periodic boundary conditions are used, however, one additional layer of fictitious cells is used in each direction having periodicity. The mesh generator will automatically initialize the necessary number of boundary cells needed to satisfy the boundary conditions.

Sec. 4A   THE FINITE-DIFFERENCE EQUATIONS

Fluid velocities and pressures are located at staggered mesh locations, u-velocities at the centers of cell sides normal to the x direction, v-velocities at the centers of cell sides normal to the y direction, and w-velocities at the centers of cell sides normal to the z direction. Pressures are at cell centers.

The finite-difference notation used corresponds to that in the *SOLA-3D* code, where half-integer index values cannot be used. The convention is that all half-integer indices are increased to the next whole integer. For example, the u-velocity located on the cell face between cells (i,j,k) and $(i+1,j,k)$ is denoted by $u_{i,j,k}^n$. A superscript n refers to the nth time step value. Thus,

$p_{i,j,k}^n$ = pressure at center of cell (i,j,k) at time level n,

$u_{i,j,k}^n$ = x direction velocity at middle of $i+1/2$ cell face at time level n,

$v_{i,j,k}^n$ = y direction velocity at middle of $j+1/2$ cell face at time level n, and

$w_{i,j,k}^n$ = z direction velocity at middle of $k+1/2$ cell face at time level n.

The basic procedure for advancing a solution through one increment in time δt consists of three steps:

1) Explicit approximations of the momentum equations 4.3 are used to compute the first guess for new time-level velocities using the initial conditions at previous time-level values for all advective, pressure, and other accelerations.
2) To satisfy the continuity equation 4.1, pressures are adjusted in each cell and the velocity changes induced by each pressure change are added to the velocities computed in step (1). An iteration is needed because the change in pressure needed in one cell to satisfy equation 4.1 will upset the balance in the six adjacent cells.
3) Finally, when there is a free surface, the surface height must be updated using equation 4.5 to give the new fluid configuration.

THE THREE-DIMENSIONAL NAVIER-STOKES MODEL Chap. 4

Repetition of these steps will advance a solution through any desired time interval. At each step, suitable boundary conditions must be imposed at all mesh and free boundary surfaces. Details of these steps and boundary conditions are given below.

A generic form for the finite-difference approximations of equations 4.3 is

$$u_{i,j,k}^{n+1} = u_{i,j,k}^{n} + \delta t \left[-\left(p_{i+1,j,k}^{n+1} - p_{i,j,k}^{n+1}\right)\middle/ \delta x_{i+1/2} \right.$$

$$\left. + g_x - FUX - FUY - FUZ + VISX + ROTX \right],$$

$$v_{i,j,k}^{n+1} = v_{i,j,k}^{n} + \delta t \left[-\left(p_{i,j+1,k}^{n+1} - p_{i,j,k}^{n+1}\right)\middle/ \delta y_{j+1/2} \right.$$

$$\left. + g_y - FVX - FVY - FVZ + VISY + ROTY \right], \text{ and}$$

$$w_{i,j,k}^{n+1} = w_{i,j,k}^{n} + \delta t \left[-\left(p_{i,j,k+1}^{n+1} - p_{i,j,k}^{n+1}\right)\middle/ \delta z_{k+1/2} \right.$$

$$\left. + g_z - FWX - FWY - FWZ + VISZ \right]. \tag{4.6}$$

The advective, viscous, and rotational acceleration terms have an obvious meaning, e.g., FUX means the advective flux of u in the x direction; $VISX, VISY$ and $VISZ$ are the component viscous accelerations; and $ROTX, ROTY$ and $ROTZ$ are the component rotational accelerations associated with a rotating coordinate system. These terms are all evaluated using old time-level (n) values for velocities. Because the pressures at time level $n+1$ are unknown at the beginning of the cycle, these equations cannot be used directly to evaluate the $n+1$ level velocities but must be combined with the continuity equation as described below. In the first step of a solution, therefore, the p^{n+1} values in these equations are replaced by p^n values to get a first guess for the new velocities.

Sec. 4A	THE FINITE-DIFFERENCE EQUATIONS

In the basic solution procedure, the specific approximations chosen for the various acceleration terms in equations 4.6 are relatively unimportant, provided they lead to a numerically stable algorithm. Special care must be exercised, however, when making approximations in a mesh with nonuniform cell sizes. Consider the approximation procedure used in the original MAC method for Cartesian coordinates. In the MAC scheme, the momentum and continuity equations 4.1 and 4.3 were combined so that the advective flux terms could be written in a divergence form (i.e., $\nabla \cdot uu$ instead of $u \cdot \nabla u$). Thus, FUX would be, for example, $\partial u^2/\partial x$ rather than $u\partial u/\partial x$. The divergence form was preferred in MAC because it provided a simple way to ensure conservation of momentum in the difference approximations.

This worked well in the original MAC scheme, which was developed for use in uniform meshes. Unfortunately, in nonuniform meshes, conservation does not automatically imply accuracy. To see this, suppose an upstream or donor-cell difference approximation is used for $FUX = \partial u^2/\partial x$, which provides a conditionally stable algorithm. The donor-cell approximation is where, e.g.,

$$FUX = (u_R <u_R> - u_L <u_L>)/\delta x_{i+1/2},$$

$$u_R = (u_{i+1,j,k} + u_{i,j,k})/2, \text{ and}$$

$$<u_R> = \begin{cases} u_{i,j,k}, & \text{if } u_{i,j,k} \geq 0; \\ u_{i+1,j,k,}, & \text{if } u_{i,j,k} < 0. \end{cases} \quad (4.7)$$

To check the accuracy of equation 4.7, we expand it in a Taylor series about the x-position, where FUX is evaluated (assuming u velocities are positive),

$$FUX = \frac{1}{2}\left(\frac{3\delta x_i + \delta x_{i+1}}{\delta x_i + \delta x_{i+1}}\right)\frac{\partial u^2}{\partial x} + 0(\delta x). \quad (4.8)$$

Clearly, the zero-order term is incorrect unless the cell widths are equal, $\delta x_i = \delta x_{i+1}$. In other words, the variable mesh reduces the order of the conservative difference approximation by one, which in this case leads to an incorrect zero-order result. If a centered, rather than a donor-cell, approximation had been used, the result would have been first-order accurate, not second order, as it is in a uniform mesh.

It does not follow from the above analysis that variable meshes are necessarily less accurate in practice because they do allow finer zoning to be used in regions where flow variables are rapidly varying. It is best, for example, to use gradual variations in cell sizes to minimize the reduction in approximation order. It is also worthwhile to look for other approximations that do not lose their formal accuracy when applied to variable meshes. In this regard, it should be noted that the reason the conservation form of the advective terms loses accuracy is because the control volumes are not centered about the positions where variables to be updated are located. The shifted-control volumes lead directly to the reduction in accuracy computed in equation 4.8.

For the *SOLA* series of codes, a modified donor-cell approximation has been developed that retains its accuracy in a variable mesh. This method approximates advective fluxes in the nonconservative form $u \cdot \nabla u$, which is necessary because of the inherent difficulty with conservative approximations noted above. In the new approximation, it is also possible to combine the donor-cell and centered-difference approximations into a single expression with a parameter β that controls the relative amount of each one. The general form of this approximation for $FUX = u\partial u/\partial x$ is

$$FUX = (u_c/\delta x_\beta)[\delta x_{i+1} DUL + \delta x_i DUR$$
$$+ \beta \text{sgn}(u_c)(\delta x_{i+1} DUL - \delta x_1 DUR)], \qquad (4.9)$$

where $u_c = u_{i,j,k}$ is the value of u at the location where FUX is evaluated, and

Sec. 4A   THE FINITE-DIFFERENCE EQUATIONS

$$DUL = (u_{i,j,k} - u_{j-1,j,k})/\delta x_i \, ,$$

$$DUR = (u_{i+1,j,k} - u_{i,j,k})/\delta x_{i+1} \, , \text{ and}$$

$$\delta x_\beta = \delta x_{i+1} + \delta x_i + \beta \text{sgn}(u_c)(\delta x_{i+1} - \delta x_i) \, .$$

The expression $\text{sgn}(u_c)$ means the sign of u_c. When $\beta = 0$, this approximation reduces to a second-order accurate, centered-difference approximation. When $\beta = 1$, the first-order, donor-cell approximation is recovered. In either case, there is no loss of formal accuracy in a variable mesh.

The basic idea underlying equation 4.9 is to weight the upstream derivative of the quantity being fluxed more than the downstream value. The weighting factors are $(1+\beta)$ and $(1-\beta)$ for the upstream and downstream derivatives, respectively. These derivatives are also weighted by cell size in such a way that the order of approximation is maintained in a variable mesh. This type of approximation scheme is used in *SOLA-3D* for all advective flux terms appearing in equation 4.6. All other acceleration terms are approximated by standard centered differences.

Continuity Equation Approximation

Velocities computed from equation 4.6 must satisfy the following discretized approximation to equation 4.1, the continuity equation

$$\begin{aligned}(u_{i,j,k} - u_{i-1,j,k})/\delta x_i &+ (v_{i,j,k} - v_{i,j-1,k})R_i/\delta y_j \\ &+ (w_{i,j,k} - w_{i,j,k-1})/\delta z_k \\ &+ 1/2\, \alpha\, (u_{i,j,k} + u_{i-1,j,k})/XC_i = 0 \, ,\end{aligned} \quad (4.10)$$

THE THREE-DIMENSIONAL NAVIER-STOKES MODEL
Chap. 4

where XC_i is the x location of the center of the ith cell. For cylindrical coordinates, $R_i = X_{IM1}/XC_i$, in which X_{IM1} is the radius (x location) of the outer edge of the last real (i.e., not boundary) cell in the mesh. In Cartesian geometry, $R_i = 1.0$ for all i.

If velocities are to satisfy equation 4.10, it is necessary to adjust the pressures and, hence, the velocities in each computational cell occupied by fluid. This is done by the following iterative process. The computational mesh is swept cell by cell starting with $i = j = k = 2$, the first nonboundary cell in the mesh. Sweeping is first carried out on i, then j, and finally on k values. When a free surface is present, the k sweep is terminated in the cell containing the free surface, which may be a different k value for each (i, j) mesh column. The pressure change needed to make velocities in cell (i, j, k) satisfy equation 4.10 is

$$\delta p = -S/(\partial S/\partial p) . \qquad (4.11)$$

In all cells below the free surface, S is the velocity divergence defined by the left side of equation 4.10. In this case, equation 4.11 is simply a Newton type of relaxation process that will produce the value of p needed to make $S = 0$. In each cell, the velocity values used in evaluating S are the most current values available during the iteration process. Using the result from equation 4.11, the new estimate for the cell pressure is

$$P_{i,j,k} + \delta p , \qquad (4.12)$$

and new estimates for the velocities located on the sides of the cell are then

$$u_{i,j,k} + \delta t \delta p / \delta x_{i+1/2} ,$$

$$u_{i-1,j,k} - \delta t \delta p / \delta x_{i-1/2} ,$$

$$v_{i,j,k} + \delta t \delta p/\delta y_{j+1/2},$$

$$v_{i,j-1,k} - \delta t \delta p/\delta y_{j-1/2},$$

$$w_{i,j,k} + \delta t \delta p/\delta z_{k+1/2}, \text{ and}$$

$$w_{i,j,k-1} - \delta t \delta p/\delta z_{k-1/2}, \tag{4.13}$$

where the velocities appearing here are also the most current values available during the iteration. To start the iteration process, the new estimated velocities from equation 4.6 are used with the pressures remaining from the previous time step.

In cells containing a free surface, a somewhat different procedure is used. The S in equation 4.11 is replaced by another expression that leads to the proper free-standing boundary conditions when driven to zero by the iteration. The boundary condition wanted is that the pressure must be a specified value, for example, p_s, at the surface. This condition is imposed by choosing the surface cell pressure $p_{i,j,k}$, such that a linear interpolation (or extrapolation) between it and the pressure in the fluid cell below the surface $p_{i,j,k-1}$ will have the correct value p_s at the actual location of the free surface. The S function giving this result is

$$S = (1-\eta)p_{i,j,k-1} + \eta p_s - p_{i,j,k}, \tag{4.14}$$

where $\eta = \delta z_c/\delta z_s$ is the ratio of the distance between the cell centers $\delta z_c = (\delta z_k + \delta z_{k+1})/2$ and the distance between the free surface and the center of cell $k-1$.

Because the surface pressure p_s is assumed constant in the code, we have arbitrarily taken it to be zero; therefore, the term in equation 4.14 involving p_s does not appear.

THE THREE-DIMENSIONAL NAVIER-STOKES MODEL
Chap. 4

A complete iteration consists of adjusting pressure and velocities in all cells containing fluid, according to equations 4.11 through equation 4.13, where S is the left side of equation 4.10 for cells full of fluid and S is given by equation 4.14 for cells containing the free surface. Convergence of the iteration is achieved when all cells have S values whose magnitudes are below some small number ϵ. Typically, ϵ is of order 10^{-3}, although it can vary with the specific problem being solved.

In some cases, convergence of the iteration may be accelerated by multiplying δp in equation 4.12 by an overrelaxation factor ω. A value of ω equal to 1.7 or 1.8 is often optimum. In no case should ω exceed 2.0; otherwise, an unstable iteration results.

Changes in the free surface height $H(x, y, t)$ are governed by the kinematic equation 4.5. Finite difference approximations to this equation are analogous to those used for momentum advection

$$H_{i,j}^{n+1} = H_{i,j}^n + \delta t(w_s - FHX - FHY), \qquad (4.15)$$

where w_s is the value of z direction velocity computed at the free surface by interpolation between the w velocities located at the top and bottom faces of the cell containing the surface. That is,

$$w_s = hw_{i,j,k-1} + (1-h)w_{i,j,k}, \qquad (4.16)$$

in which $h = (z_k - H_{i,j})/\delta z_k$, where z_k is the height of the top of the surface cell having index k. FHX and FHY are the horizontal advective fluxes of H and are computed by expressions similar to equation 4.9,

$$FHX = (u_c/\delta x_\beta)[\delta x_R DHL + \delta x_L DHR$$
$$+ \beta sgn(u_c)(\delta x_R DHL - \delta x_L DHR)], \qquad (4.17)$$

Sec. 4A   THE FINITE-DIFFERENCE EQUATIONS

where

$$u_c = (u_{i,j,k} + u_{i-1,j,k})/2 \,,$$

$$\delta x_R = (\delta x_{i+1} + \delta x_i)/2 \,,$$

$$\delta x_L = (\delta x_i + \delta x_{i-1})/2 \,,$$

$$\delta x_\beta = \delta x_R + \delta x_L + \beta \operatorname{sgn}(u_c)(\delta x_R - \delta x_L) \,,$$

$$DHL = (H_{i,j} - H_{i-1,j})/\delta x_L \,, \text{ and}$$

$$DHR = (H_{i+1,j} - H_{i,j})/\delta x_R \,.$$

To perform the above calculations, it is necessary to know the k index of the cell containing the free surface. This index is also required to limit the k values considered in all other computations. For efficiency, the program keeps these k values in an array $KT(i,j)$ that must be updated along with the H values.

The free surface has been defined by $z = H(x,y,t)$, but this does not mean the technique applies only to horizontal surfaces. For example, liquid flow down inclined surfaces can be investigated by using a vector body acceleration (g_x, g_y, g_z) that is directed at the desired angle to the $z = 0$ plane of the mesh. Also, instability of inverted surfaces can be studied by simply changing the sign of g_z.

Relatively small modifications are required to convert the free surface to a rigid, curved surface or to add surface tension by defining a surface pressure p_s in terms of local surface curvatures.

In addition to the free surface pressure boundary conditions, it is also necessary to set conditions at all mesh boundaries and at surfaces of all internal obstacles.

At the mesh boundaries, a variety of conditions may be set using the layer of fictitious cells surrounding the mesh. Consider, for example, the boundary separating the $i = 1$ and $i = 2$ layer of cells.

THE THREE-DIMENSIONAL NAVIER-STOKES MODEL   Chap. 4

The $i = 1$ cells are fictitious in the sense that variable values are set in these cells to satisfy boundary conditions rather than calculated according to the finite-difference expressions used at all interior mesh locations. If this boundary is to be a rigid wall, the normal velocity there must be zero. The tangential velocity can either be set to zero for a no-slip type of wall, or the normal derivative of the tangential velocity can be set to zero corresponding to a free-slip boundary condition. Thus, for a free-slip, rigid wall, the boundary conditions are, for all j, k

$$u_{1,j,k} = 0.0 ,$$

$$v_{1,j,k} = v_{2,j,k} ,$$

$$w_{1,j,k} = w_{2,j,k} ,$$

$$p_{1,j,k} = p_{2,j,k} , \text{ and}$$

$$H_{1,j} = H_{2,j} . \tag{4.18}$$

For a no-slip wall, the conditions are the same except for u and w, which must be replaced by

$$v_{1,j,k} = -v_{2,j,k} , \text{ and}$$

$$w_{1,j,k} = -w_{2,j,k} . \tag{4.19}$$

These conditions are imposed on the velocities obtained from the momentum equations and after each pass through the mesh during the pressure iteration.

For a specified in or out flow boundary, the special values for u, v, and w should be set in the $i = 1$ layer of cells. It should also be noted that specified in or out flow can be different at each j, k (i.e., y, z) location. This allows an extremely wide range of possible boundary conditions.

Continuative outflow boundaries, where fluid is to flow smoothly out of the mesh causing no upstream effects, are always a problem for incompressible flow calculations. Potentially, any prescription

Sec. 4A   THE FINITE-DIFFERENCE EQUATIONS

can affect the entire flow field. In *SOLA-3D*, the continuative boundary conditions used at the $i = 1$ boundary are the vanishing of all normal derivatives, that is for all j, k

$$u_{1,j,k} = u_{2,j,k},$$

$$v_{1,j,k} = v_{2,j,k},$$

$$w_{1,j,k} = w_{2,j,k},$$

$$p_{1,j,k} = p_{2,j,k}, \text{ and}$$

$$H_{1,j} = H_{2,j}. \tag{4.20}$$

These conditions, however, are only imposed after applying the momentum equations and not after each pass through the pressure iteration. Omitting these conditions during the pressure adjustment provides additional "softness" at the boundary that helps to minimize upstream effects.

For periodic boundary conditions in the x direction, both the $i = 1$ and $i = IMAX$ boundaries must be set to reflect the periodicity. This is most easily done when the period length is chosen equal to the distance from the boundary between the $i = 1$ and 2 cells to the boundary between the $i = IMAX - 2$ and $IMAX - 1$ cells. That is, two layers of cells, $i = IMAX$ and $i = IMAX - 1$, are reserved for setting periodic boundary conditions at the far side of the mesh. The conditions for periodic flow in the x direction are then, at $i = 1$ for all j, k

$$u_{1,j,k} = u_{IM2,j,k},$$

$$v_{1,j,k} = v_{IM2,j,k},$$

$$w_{1,j,k} = w_{IM2,j,k}, \text{ and}$$

$$H_{1,j} = H_{IM2,j}, \tag{4.21a}$$

THE THREE-DIMENSIONAL NAVIER-STOKES MODEL Chap. 4

and at $IM1$ or $IMAX$ for all j,k

$$u_{IM1,j,k} = u_{2,j,k},$$

$$v_{IM1,j,k} = v_{2,j,k},$$

$$w_{IM1,j,k} = w_{2,j,k},$$

$$p_{IM1,j,k} = p_{2,j,k},$$

$$(p_s)_{IM1,j,k} = (p_s)_{2,j,k},$$

$$H_{IMAX,j} = H_{3,j},$$

$$v_{IMAX,j,k} = v_{3,j,k}, \text{ and}$$

$$w_{IMAX,j,k} = w_{3,j,k}, \tag{4.21b}$$

where $IM1 = IMAX - 1$ and $IM2 = IMAX - 2$. In this case, these conditions are imposed on velocities computed from the momentum equations and after each pressure iteration.

At constant pressure, boundary conditions at the $i = 1$ boundary are set by keeping the pressures in the $i = 2$ layer of cells constant and otherwise treating the boundary as continuative.

Boundary conditions similar to these are used at all other mesh boundaries. Of course, the appropriate normal and tangential velocities must be used in each case.

Sec. 4A   THE FINITE-DIFFERENCE EQUATIONS

The normal stress, or specified pressure, condition at a free surface is satisfied by the pressure iteration using equation 4.14. In addition, tangential stress conditions must be specified at a free surface by defining velocity components immediately outside the surface. These values are needed in the finite-difference expressions used for mesh points at the surface. The specifications used in *SOLA-3D* are identical to those used in many earlier *MAC* codes. The velocity component is set at the top face of each cell containing the free surface, such that the velocity divergence in the surface cell will be zero. If the surface cell has an empty neighbor at $i \pm 1$ or $j \pm 1$, then the normal velocity components at those faces of the cell are set equal to the corresponding velocities at $k-1$. At level $k+1$, the u and v velocity components are set equal to the components at level k. The simplicity of these conditions results from the limitation that surface slopes do not exceed the cell aspect ratios $\delta z/\delta x$ or $\delta z/\delta y$.

The definition of obstacles within a mesh is accomplished by flagging those mesh cells that define the obstacles. Flag values must be programmed for each application. No velocities or pressures are computed in obstacle cells, and all velocity components on faces of obstacle cells are automatically set to zero.

It should also be noted that because all velocity components within obstacles are set to zero, no-slip tangential velocity conditions are approximated at their surfaces but are only first order accurate. That is, tangential velocities are zero at locations shifted into the obstacles one half-cell width from the actual boundary location.

There are several restrictions on the time-step size that must be observed to avoid numerical instabilities. If requested, the *SOLA-3D* code will automatically adjust the time step to be as large as possible without violating the stability conditions. It will also reduce the time step when pressure iterations exceed a normal value of 50 per cycle. Generally, the time step will move up or down with a 5 percent change per cycle, unless a stability condition is violated; in such a case, a large reduction may occur.

If the automatic time-step control is not requested, time steps must satisfy the following criteria. First, fluid must not be permitted to flow across more than one computational cell in one time step,

THE THREE-DIMENSIONAL NAVIER-STOKES MODEL
Chap. 4

$$\delta t < \min \{\delta x_i/|u|, \delta y_j/|v|, \delta z_k/|w|\} . \tag{4.22}$$

This is clearly an accuracy condition because the advective fluxes are approximated by difference assumptions that only use information from neighboring cells. However, it is also a stability condition, as can be verified by performing a linear stability analysis on the advective portion of the momentum equation approximations. In fact, a review of worst case situations indicates that the minimum value in the previous test (4.22) should be multiplied by 0.25, which allows a stability safety margin for most applications.

$$\delta t < 0.25 \min \{\delta x_i/|u|, \delta y_j/|v|, \delta z_k/|w|\} . \tag{4.22a}$$

A linear analysis indicates that the time step, additionally, must be limited when a nonzero value of kinematic viscosity is used. This condition,

$$\nu \delta t < 0.5/\max \{1/\delta x_i^2, 1/\delta y_j^2, 1/\delta z_k^2\} , \tag{4.23}$$

roughly restricts momentum to not diffuse more than one mesh cell in one time step.

One more parameter, $ALPHA$, must also be set correctly to ensure stability. This parameter controls the relative amounts of donor-cell and centered differencing used for the momentum advection terms. When $ALPHA = 1.0$ is used, the above stability conditions are sufficient. Generally, one should always use an $ALPHA$ value such that

$$1.0 \geq ALPHA > \delta t \max \{|u|/\delta x_i, |v|/\delta y_j, |w|/\delta z_k\} . \tag{4.24}$$

Sec. 4A   THE FINITE-DIFFERENCE EQUATIONS

For instance, if condition (4.22a) is satisfied, then we must have $ALPHA > 0.25$. In no case should $ALPHA$ exceed 1.0.

4B. Application to Tsunami Wave Formation

The first application of the incompressible Navier-Stokes model to tsunami wave formation was by Garcia[4] using a two-dimensional arbitrary boundary marker and cell technique to study tsunamis in the vicinity of their source. The method was applied to the Mendocino Escarpment for hypothetical ocean floor displacements of 10 meters in a few seconds to a minute, which might result from a major earthquake on the San Andreas fault. The wave was followed numerically for the first 200 sec. A single hump was first generated that split into two crests moving in opposite directions at a speed slightly less than shallow water wave speed. The waves were slightly dispersive and had a crest elevation above mean water level of about 1 meter and a period of about 1.5 minute. The period of major tsunamis measured in the Pacific Ocean is generally from 10 to 30 minutes. The periods are estimated from tide gauge records assuming that the period remains nearly constant during the shoaling of the wave from the deep ocean. The wave height of major tsunamis is generally given at 1 ± 0.5 meter and is estimated from the tide gauge records assuming various approximate models for the wave height growth and extrapolating back to the deep ocean.

The first application of the three-dimensional incompressible Navier-Stokes model to tsunami formation was by Mader, Tangora, and Nichols,[5] and by Mader.[6]

The Hawaiian tsunami of November 29, 1975, has been investigated by Loomis. He described the observed runup heights in reference 7 and a numerical study of the tsunami source in references 8 and 9.

The tsunami was generated by an earthquake with a magnitude of 7.2 on the Richter scale, near the Hawaii Volcanoes National Park. Near the source, the first wave was smaller than the

THE THREE-DIMENSIONAL NAVIER-STOKES MODEL Chap. 4

second. Coincident with the earthquake was considerable subsidence (up to 3 meters) of the shoreline.

Loomis, in reference 8, examined a model of the southeastern coast of Hawaii. The bottom slopes seaward at a ratio of 1:15 until it reaches a constant depth of 6000 meters. The sources examined by Loomis included both initial uplifts and depressions, and he reported that such source motions would not generate the essential features of the tsunami; that is, a second wave larger than the first.

In addition to the sources studied by Loomis, a landslide source was modeled in reference 6. The landslide model has been evaluated by Cox in reference 10. He concluded that a landslide could not be distinguished from strictly tectonic displacement by the comparison of arrival times and travel times.

The *SWAN* code described in Chapter 2 was used to solve the shallow water, long wave equations and to examine the tsunami generation problem. The *SWAN* results confirmed Loomis's calculated results reported in reference 8.

The *SOLA-3D* code, which solves the three-dimensional incompressible Navier-Stokes equations, was also used to model the tsunami.

The Calculated Shallow Water Wave Results

The model was identical to that described by Loomis in reference 7. A 40 by 69 rectangular region of 207 kilometers along the coast and 120 kilometers seaward is described using a mesh of 3 kilometers by 3 kilometers. The bottom slopes at a ratio of 1:15 until it reaches a depth of 6 kilometers. The source is a bottom slope subsidence programmed to vary with time. The source is 30 kilometers wide, of which half is included in the calculation and is separated from the other half by a reflective boundary, as shown in Figure 4.1.

The calculated wave profile is shown at various locations along the shoreline as a function of time in Figure 4.2 for a source displacement of 3 meters, and in Figure 4.3 for a source displacement of 1 meter, followed by an additional 2 meters displacement 10 minutes later. Surface profiles are shown in Figure 4.4.

Sec. 4B APPLICATION TO TSUNAMI WAVE FORMATION

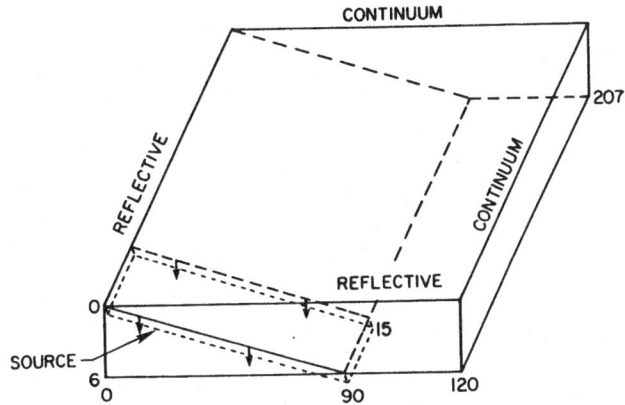

Fig. 4.1. Sketch of model used to numerically simulate the tsunami generation.

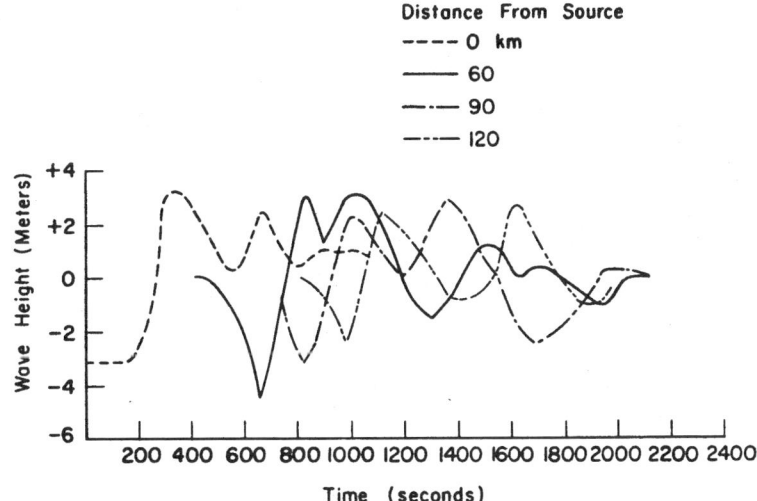

Fig. 4.2. Shoreline wave heights for shallow water wave model resulting from an initial source displacement of 3 meters.

The observed larger second wave can be reproduced by a source that has its displacement change with time. Such a possibility was suggested by Ando, who suggested that the earthquake was a rather

THE THREE-DIMENSIONAL NAVIER-STOKES MODEL
Chap. 4

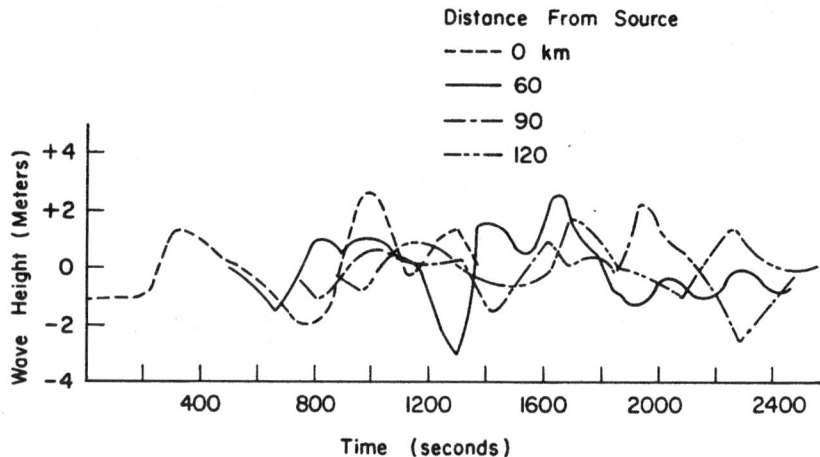

Fig. 4.3. Shoreline wave heights for shallow water wave model resulting from an initial source displacement of 1 meter, followed by an additional 2 meters displacement 10 minutes later.

slow rupture lasting 100 sec; however, the displacement change required by the model is of longer duration and includes two fast ruptures.

Another source investigated was an undersea landslide. The landslide ocean bottom profile assumed the bottom dropped 3 meters at the shoreline and slid to form the profile shown in Figure 4.5. Landslides are observed to pile up at the bottom one-third of their run. This gives the surface wave profile shown in Figure 4.6. The calculations were performed on the University of Hawaii Harris computer using the Hawaii version of the *SWAN* code.

The shoreline wave heights at various times for the shallow water wave model with the initial water surface displacement of Figure 4.6 are shown in Figure 4.7. Although the wave heights are consistent with the observed behavior of the tsunami, we must check the results with the *SOLA* code since it was demonstrated in Chapter 3 that the shallow water model is inadequate to describe the waves generated from surface deformations of the water surface.

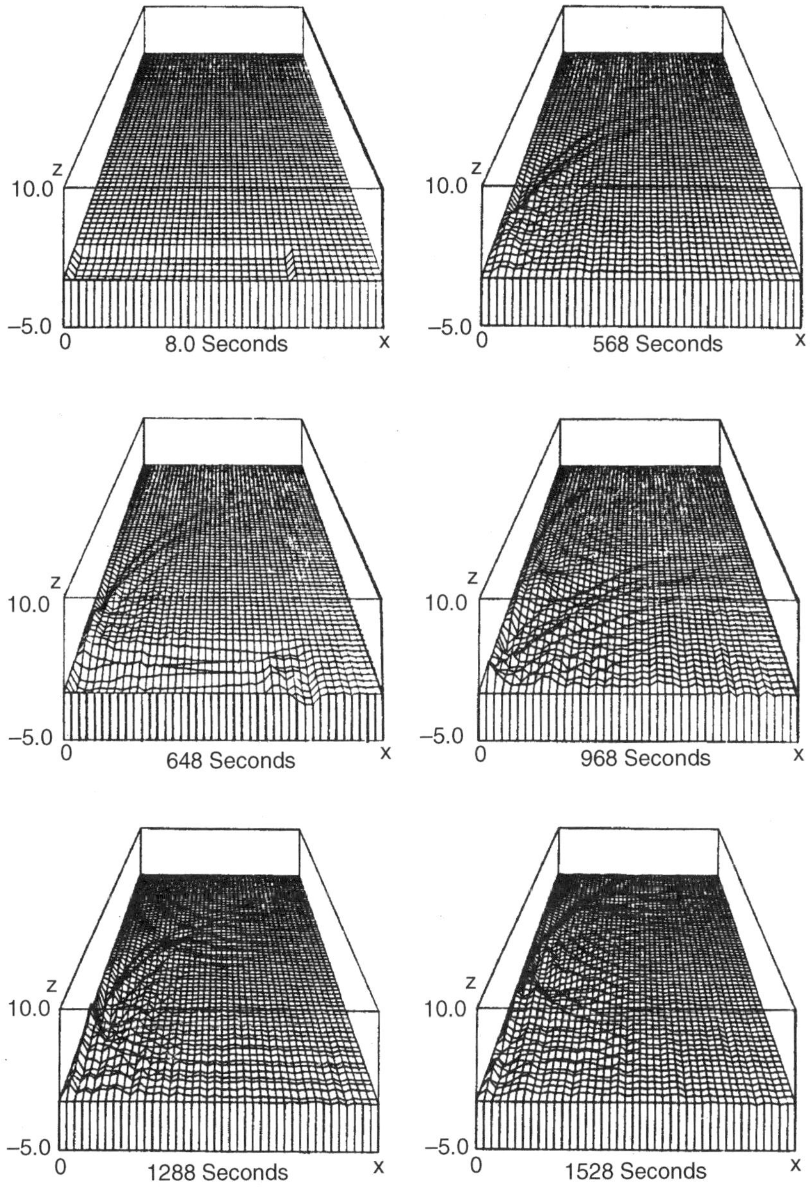

Sec. 4B APPLICATION TO TSUNAMI WAVE FORMATION

Fig. 4.4. Surface profiles for linear shallow water wave model as a function of time, resulting from an initial shore displacement of 1 meter, followed by an additional 2 meters displacement 10 minutes later.

THE THREE-DIMENSIONAL NAVIER-STOKES MODEL

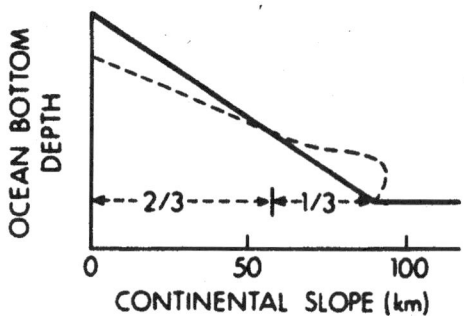

Fig. 4.5. Sketch of the final ocean bottom profile after a landslide for the source region.

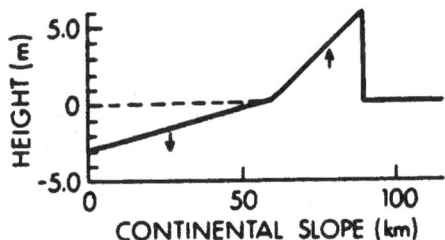

Fig. 4.6. Sketch of height of water surface after a landslide on the ocean bottom.

The Calculated Navier-Stokes Results

The geometry of the model used to calculate the tsunami is shown in Figure 4.1. The mesh used in the calculation had 20 cells in the x direction, 25 cells in the y direction, and 18 cells in the z direction. The 20 cells in the x direction were 6 kilometers wide. The 18 cells in the z direction starting at the ocean floor were 100 meters high for the first two cells, and 400 meters thereafter. The water depth was 6000 meters and the surface was located at the center of cell 17. The 25 cells in the y direction starting at the source were 3.0 kilometers for the first 5 cells that described the source (15 kilometers wide). The remaining cell widths were 5.75, 6.16, 6.56, 6.97, 7.37, 7.78, 8.18, 8.59, 9.0, 9.4, 9.8, 10.2, 10.6, 11.0, 11.4, 11.8, 12.2, 12.6, 13.0, and 13.5 kilometers, for a total of 206.8 kilometers.

Sec. 4B APPLICATION TO TSUNAMI WAVE FORMATION

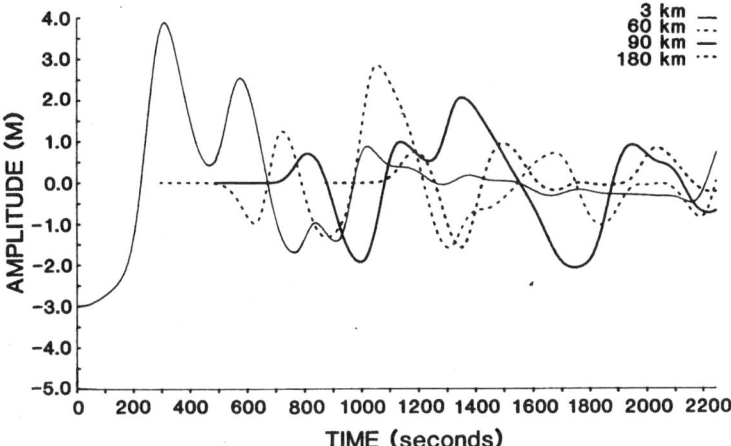

Fig. 4.7. Shoreline wave heights for a shallow water wave model resulting from the initial water surface displacement shown in Figure 4.6.

The viscosity coefficient was 2.0 g-sec^{-1} meter^{-1} (0.02 poise). The gravity constant g_z was -9.8 meters sec^{-2}, and g_x and g_y were 0.0. The time step for the calculation was 5 sec. The tsunami source was modeled by a 3 meters deep depression or elevation of 90 by 15 kilometers of the water surface, as shown in Figure 4.1.

The calculated wave profiles are shown at various locations along the shoreline as a function of time in Figures 4.8, for a source of 3 meters depression of the water surface. The calculation for a 3 meters uplift gave mirror images of the profiles for the 3 meters depression of the water surface. Surface profiles are shown in Figure 4.9.

The calculated wave profiles are shown in Figure 4.10 at various locations along the shoreline as a function of time for a landslide source. The observed tsunami wave profile of the 1975 Hawaiian tsunami, near the source of the second wave larger than the first, is not reproduced by a landslide source in an incompressible three-dimensional Navier-Stokes calculation. This contrasts with results obtained using the shallow water model.

159

THE THREE-DIMENSIONAL NAVIER-STOKES MODEL

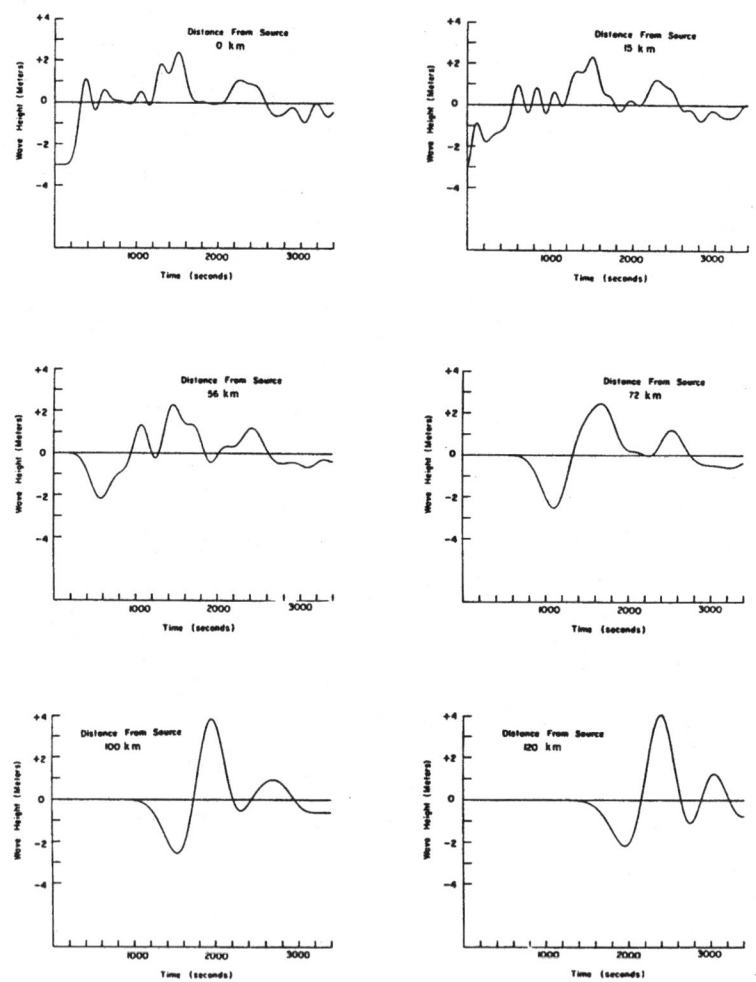

Fig. 4.8. Shoreline wave heights for the three-dimensional incompressible Navier-Stokes model as a function of time resulting from an initial source of a 3 meters depression of the water surface.

Sec. 4B APPLICATION TO TSUNAMI WAVE FORMATION

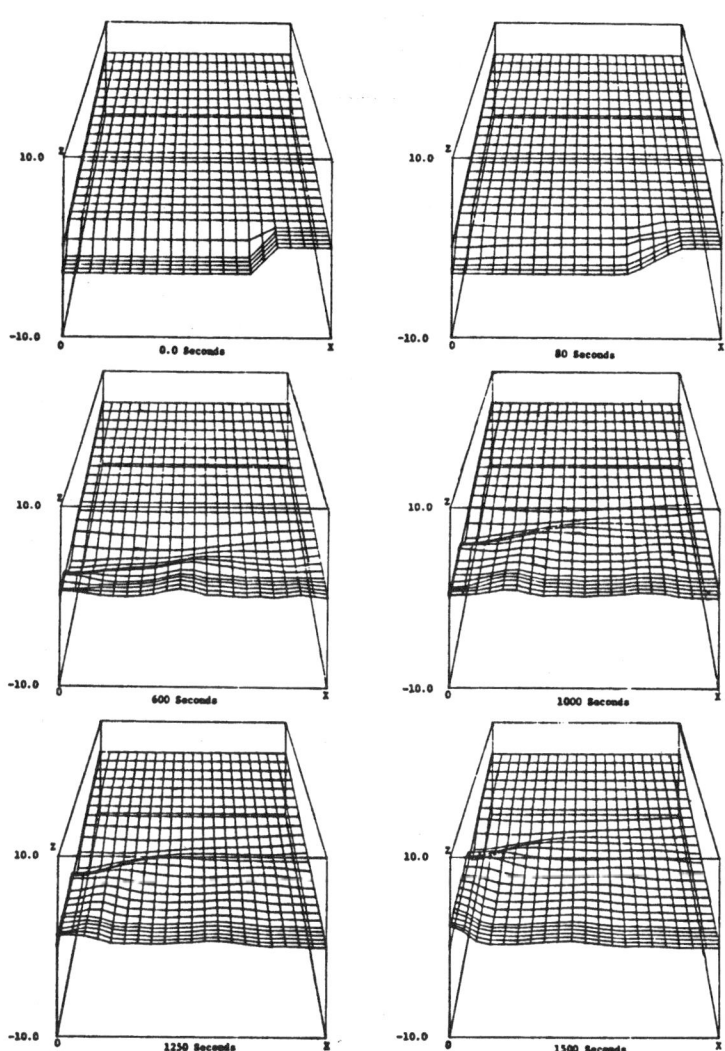

Fig. 4.9. Surface profiles for three-dimensional incompressible Navier-Stokes model as a function of time resulting from an initial source of a 3 meters depression of the water surface.

THE THREE-DIMENSIONAL NAVIER-STOKES MODEL Chap. 4

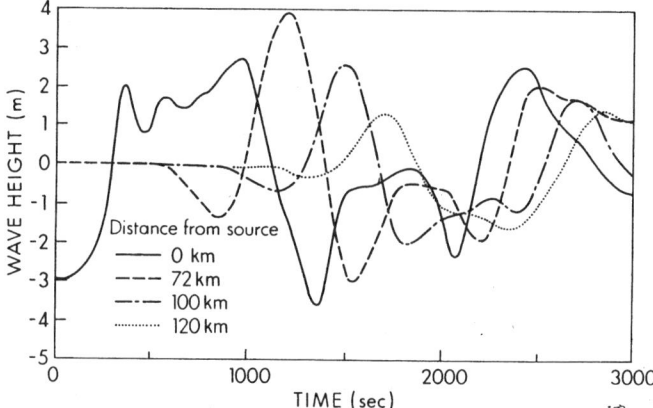

Fig. 4.10. Shoreline wave heights for a three-dimensional incompressible Navier-Stokes model calculation with the landslide source shown in Figure 4.6.

A source with a 3 meters uplift of the water surface was consistent with the observed tsunami wave profile. These calculations do not support a landslide source for the 1975 Hawaii tsunami.

Conclusions

The differences between the shallow water and full Navier-Stokes calculations are that the water waves formed in the full Navier-Stokes calculations are deep water waves that move slower than the shallow water waves formed in the shallow water calculations. The nature of the surface collapse is also different, with the collapse occurring throughout the source region in the Navier-Stokes calculations and mostly at the sides in the shallow water calculations.

The observed tsunami wave profile of the 1975 Hawaii tsunami, of the second wave being larger than the first wave near the source, is not reproduced by a landslide source, but it is reproduced by a simple uplift or drop of the water surface over the source area.

The shallow water approximation is not appropriate for studying waves generated from surface deformations that are small, relative to the water depth. In the next chapter we will investigate the limitations of the shallow water approximation for modeling tsunami waves.

4C. The 1994 Skagway Tsunami

The Skagway tsunami of November 3, 1994 is described in Section 2G of Chapter 2. A tsunami wave with a period of about 3 minutes and a maximum height of 30 feet occurred in the Taiya Inlet at Skagway, Alaska. The tsunami was a result of a massive underwater landslide of sediment deposited by the Skagway River. The tsunami was modeled in Chapter 2 using the shallow water *SWAN* code.

The Skagway tsunami was also modeled using the *SOLA-3D* code. A single slide model for the Skagway tsunami, that closely approximated the more realistic 3-slide Landslide model, was used for the three-dimensional *SOLA-3D* calculation. A *SWAN* calculation for the same single slide model was also performed for comparision. The *SOLA-3D* FORTRAN code and input file used to model the Skagway tsunami and the computer movies are on the NMWW CD-ROM in the directory /TSUNAMI.MVE/SOLA/SKAGWAY. Since the wave length and period of the resulting tsunami is large compared to the shallow depth of the inlet, the tsunami wave is adequately approximated by a shallow water wave. The tsunami wave profiles generated by the *SOLA-3D* code and the *SWAN* code were similar in amplitude and period.

References

1. C. W. Hirt, B. D. Nichols, and N. C. Romero, "*SOLA* — A Numerical Solution Algorithm for Transient Fluid Flows," Los Alamos National Laboratory report LA–5852 (1975).
2. C. W. Hirt, B. D. Nichols, and N. C. Romero, "*SOLA* — A Numerical Solution Algorithm for Transient Fluid Flows – Addendum," Los Alamos National Laboratory report LA–5852, Add. (1976).
3. C. W. Hirt, B. D. Nichols, and L. R. Stein, "*SOLA-3D* — A Numerical Solution Algorithm for Transient Three-Dimensional Flows," Los Alamos National Laboratory, Group T-3, unpublished internal report (1985).
4. W. J. Garcia, "A Study of Water Waves Generated by Tectonic Displacements," College of Engineering, University of California at Berkeley report HEL 16–9 (1972).

5. Charles L. Mader, Robert E. Tangora, and B. D. Nichols, "A Model of the 1975 Hawaii Tsunami," Natural Science of Hazards — The International Journal of the Tsunami Society, pp. C1–C8 (1982).
6. Charles L. Mader, "A Landslide Model for the 1975 Hawaii Tsunami," Science of Tsunami Hazards, Vol. 2, No. 2, pp. 71–77 (1984).
7. Harold G. Loomis, "The Tsunami of November 29, 1975, in Hawaii," Hawaii Institute of Geophysics report HIG–75–21 (1975).
8. Harold G. Loomis, "On Defining the Source of the 1975 Tsunami in Hawaii," JIMAR report to Nuclear Regulatory Commission (1978).
9. M. A. Sklarz, L. Q. Spielvogel, and H. G. Loomis, "Numerical Simulation of the November 29, 1975, Island of Hawaii Tsunami by the Finite Element Method," Journal of Physical Oceanography, Vol. 9, No. 5, pp. 1022–1031 (1979).
10. Doak C. Cox, "Source of the Tsunami Associated with the Kalapana (Hawaii) Earthquake of November 1975," Hawaii Institute of Geophysics report HIG–80–8 (1980).

5

EVALUATION OF INCOMPRESS-IBLE MODELS FOR MODELING WATER WAVES

The shallow water code *SWAN* and the incompressible Navier-Stokes code *ZUNI* have been used to model the development of a tsunami wave from an initial sea surface displacement, the propagation of a tsunami wave, and the resulting shoaling and flooding. The generation and propagation of a tsunami wave has also been modeled using a linear gravity model. The discrepancies between the shallow water and the more realistic Navier-Stokes and linear gravity models are quantified. These studies were performed by the author at the University of Hawaii Joint Institute for Marine and Atmospheric Research in the late 1980s and were published in references 1, 2 and 3.

5A. Tsunami Wave Generation

Initial Surface Displacement Study – *SWAN* Model

A 1 meter high Airy half-wave surface displacement with a width of 45 kilometers in 4550 meters deep water was studied. This approximates within cell resolution a Gaussian wave with a Gaussian break width of 10 kilometers. Calculations were performed using a 250 meters wide mesh of 400 by 4 cells and at 0.5 sec intervals.

The wave height in meters as a function of distance is shown in Figure 5.1 at various times. The initial surface displacement separates into two shallow water waves with a height of 0.5 meter and length of 45 kilometers. At maximum height the vertical velocity at the center of the wave is 0.0232 meter/sec. An Airy wave with a 0.5 meter half width, 90 kilometers wave length in 4450 meters of water has a vertical velocity of 0.0235 to 0.0226 meter/sec at maximum height. The wave speed is 208.36 and the group velocity is 202.89 meters/sec. The shallow water approximation used in the *SWAN* code of constant vertical velocity introduces errors of about 5 percent.

EVALUATION OF INCOMPRESSIBLE MODELS FOR MODELING WATER WAVES

Chap. 5

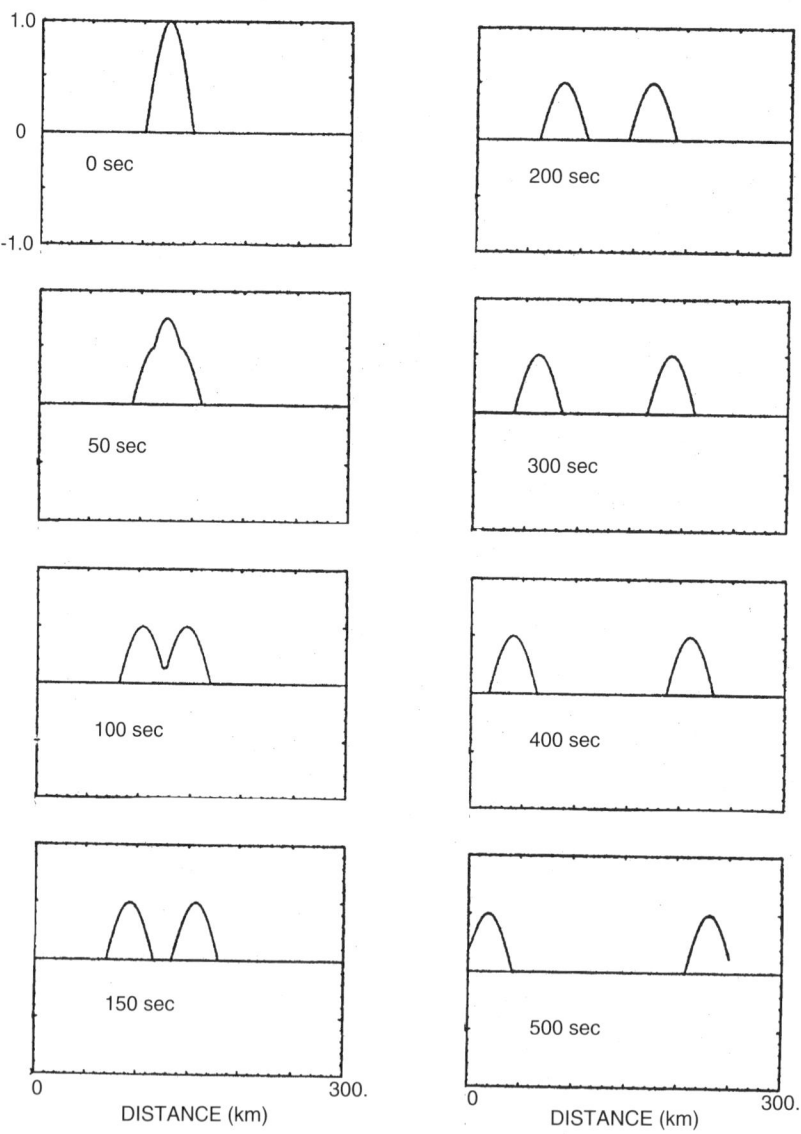

Fig. 5.1. The wave height in meters as a function of distance at various times for a 1 meter high surface displacement with a width of 45 kilometers in 4550 meters deep water for the nonlinear shallow water model using the *SWAN* code.

Sec. 5A TSUNAMI WAVE GENERATION

The nonlinear feature of the shallow water model included in the *SWAN* code results in a small trailing wave with an amplitude of less than 0.01 meter.

Initial Surface Displacement Study – *ZUNI* Model

An approximately 1 meter high (within cell resolution) Gaussian surface water displacement with a Gaussian break width of 10 kilometers (which is equivalent to an Airy wave with a half-wave length of 45 kilometers) in 4550 meter deep water was studied. The calculations for this geometry were performed with 15 cells in the Y or depth direction and 68 cells in the X or distance direction. The cells were rectangles 450 meters high in the Y direction and 4000 meters long in the X direction. The time increment was 3 sec. The water level was placed at 4550 meters or 50 meters up into the eleventh cell. The viscosity coefficient used was 0.02 poise, a value representative of the actual viscosity for water. The viscosity did not significantly affect the results.

The wave height in meters as a function of distance is shown in Figure 5.2 at various times. The initial surface displacement separates into two nearly shallow water waves with a height of 0.5 meter and length of 45 kilometers. At 100 sec two waves have formed. The vertical velocity at the top center of the wave is 0.0231 meter/sec and 0.2158 meter/sec near the bottom of the wave. An Airy wave with a 0.5 meter half width, 90 kilometers wave length in 4450 meters of water has a center vertical velocity of 0.0235 to 0.0226 meter/sec. The Airy wave speed is 208.36 and the group velocity is 202.89 meters/sec.

As the wave propagates, the wave height decreases, the slope of the front of the wave becomes less, and small waves form behind the main wave. After the wave has propagated for 500 sec, the wave height has decreased to 0.445 meter, and the vertical velocity at the center of the wave has decreased to 0.0217 meter/sec near the peak of the wave to 0.0203 at the bottom of the wave. A train of waves has developed behind the main waves with maximum negative amplitude of 0.1 meter and positive amplitude of 0.05 meter.

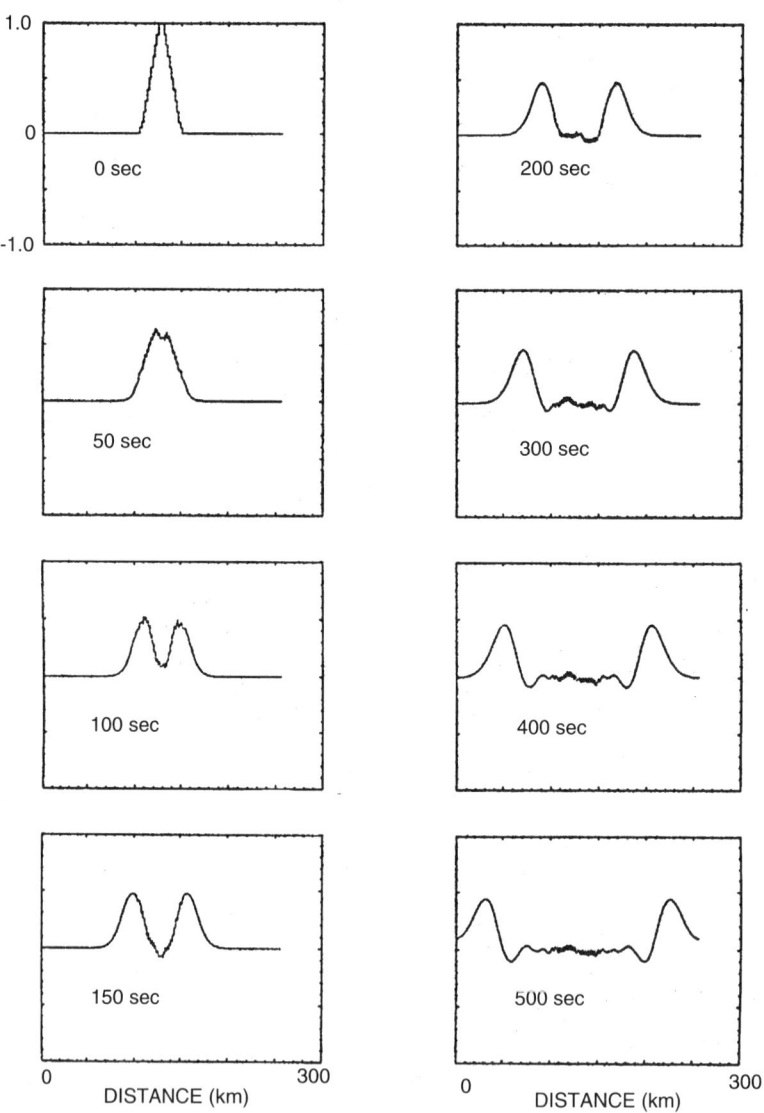

Fig. 5.2. The wave height in meters as a function of distance at various times for a 1 meter high surface displacement with a width of 45 kilometers in 4550 meters deep water for the Navier-Stokes water wave model using the *ZUNI* code.

Sec. 5A					TSUNAMI WAVE GENERATION

Initial Surface Displacement Study – *LGW* Model

The analytical methods for solving the linear gravity model were described by Professor George Carrier[4,5]. The two-dimensional linear gravity wave with a Gaussian, a square wave, or a time dependent Gaussian displacement is solved using Fourier transforms by the *LGW* code for any time of interest.

The wave description is obtained for any uniform depth, density, gravity and Gaussian break width or square wave half width. The wave height, vertical and horizontial velocities, and pressure are calculated at any depths desired. The distance scale is chosen to be a tenth of a Gaussian break width so that a wave half-width is described by about 50 space increments. Since the model is symmetrical about the center of the initial displacement, only half of the wave profile is calculated. The *LGW code* is included on the NMWW CD-ROM.

A 1 meter high Gaussian surface water displacement with a Gaussian break width of 10 kilometers (which is equivalent to an Airy wave with a half-wave length of 45 kilometers) in 4550 meters deep water was studied. Since the model is symmetrical about the center of the initial displacement, only half of the wave is modeled analytically. The analytical linear gravity wave calculations were performed using 256 points with a thickness of 1000 meters. The wave profile was calculated for each time selected.

The wave height in meters as a function of distance is shown in Figure 5.3 at various times. The initial surface displacement separates into two nearly shallow water waves with a height of 0.5 meter and length of 45 kilometers. At 100 sec the two waves have formed and the vertical velocity at the center of the wave is 0.0238 meter/sec at the top of the wave to 0.217 meter/sec near the bottom of the wave. An Airy wave with a 0.5 meter half width, 90 kilometers wave length in 4450 meters of water has a vertical velocity of 0.0235 to 0.0226 meter/sec at the center of the wave. The Airy wave speed is 208.36 and the group velocity is 202.89 meters/sec.

EVALUATION OF INCOMPRESSIBLE MODELS FOR MODELING WATER WAVES

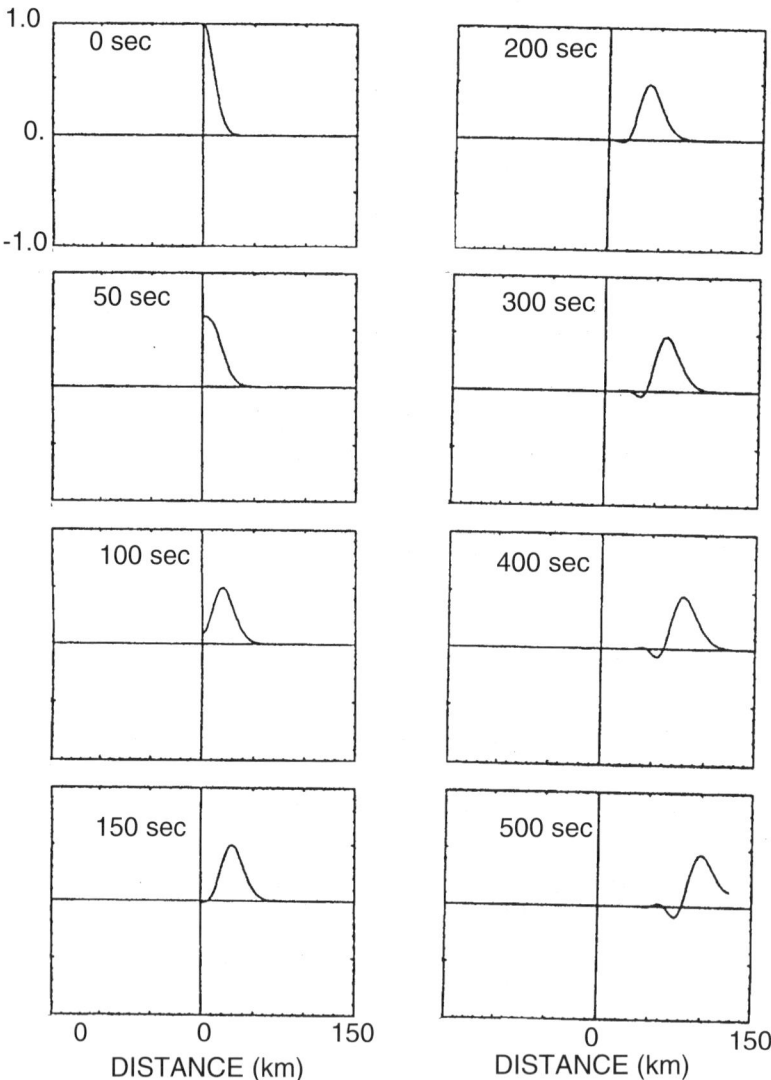

Fig. 5.3. The wave height in meters as a function of distance at various times for a 1 meter high Gaussian displacement with a Gaussian break length of 10 kilometers (equivalent to an Airy half wave length of 45 kilometers) in 4550 meters deep water for the linear gravity water wave model using the LGW code.

Sec. 5A TSUNAMI WAVE GENERATION

As the wave propagates, the wave height decreases, the slope of the front of the wave becomes less, and small waves form behind the main wave. After the wave has propagated for 500 sec, the wave height has decreased to 0.4567 meter, and the vertical velocity at the center of the wave has decreased to 0.0216 meter/sec near the peak of the wave to 0.02015 at the bottom of the wave. A train of waves has developed behind the main waves with maximum negative amplitude of 0.1 meter and positive amplitude of 0.03 meter.

Comparisions

In Table 5.1 are listed for comparision the peak wave height and the associated vertical velocity at the top and bottom of the wave for various times for the nonlinear shallow water, the full Navier-Stokes and the linear gravity wave models. The initial surface displacement was chosen to be as similar as possible for each of the models. The initial surface displacement was chosen to be characteristic of a typical earthquake generated tsunami wave at its source.

The surface displacement was one meter high spread over 45 kilometers in 4550 meters of water for a width-to-depth ratio of about 10. The initial surface displacement resulted in two waves traveling in opposite directions with half the initial height.

The nonlinear aspect of the shallow water wave appeared as very small trailing waves. The dispersion aspects of the linear gravity wave and the Navier-Stokes waves resulted in the front of the wave spreading, along with a corresponding decrease in the slope of the front of the wave and the generation of a following train of waves with amplitudes of a tenth or less of the peak wave height after the wave had traveled 105 kilometers in 500 sec (210 meters/sec). The peak wave height of the Navier-Stokes wave was lower than for the linear gravity wave. However, this difference was an order of magnitude less than the difference between either wave and the shallow water wave.

EVALUATION OF INCOMPRESSIBLE MODELS FOR MODELING WATER WAVES

Chap. 5

TABLE 5.1

A 1 m High, 48 Kilometers Wide Source

Time sec	Height meters	Bottom Velocity meters/sec	Top Velocity meters/sec
Navier-Stokes Wave -*ZUNI* Code			
0.	1.000	0.0000	0.0000
50.		0.0195	0.0202
100.	0.493	0.0216	0.0231
200.	0.475	0.0216	0.0233
300.	0.464	0.0211	0.0227
400.	0.459	0.0208	0.0224
500.	0.445	0.0203	0.0217
Linear Gravity Wave -*LGW* Code			
0.	1.000	0.0000	0.0000
50.		0.0192	0.0220
100.	0.496	0.0217	0.0238
200.	0.489	0.0214	0.0234
300.	0.478	0.0211	0.0228
400.	0.467	0.0206	0.0222
500.	0.457	0.0201	0.0216
Shallow Water Wave -*SWAN* Code			
0.	1.000	0.0000	0.0000
50.		0.0231	0.0231
100.	0.500	0.0232	0.0232
200.	0.499	0.0232	0.0232
300.	0.500	0.0232	0.0232
400.	0.499	0.0232	0.0232
500.	0.500	0.0232	0.0232

Sec. 5A	TSUNAMI WAVE GENERATION

The differences described will become less with increasing wave length-to-depth ratios and become greater for decreasing wave length-to-depth ratios. To quantify this effect we used the linear gravity wave model to examine the propagation of waves of various widths-to-depths ratios.

Initial Surface Displacement Width Study-LGW Model

The linear gravity wave model was used to investigate the waves formed from initial surface displacements of width-to-depth ratios of 40 to 0.5 (Gaussian break widths of 40 to 0.5 kilometers) in 4550 meters deep water. The wave height after the wave had travel ten times its initial width for depth ratios of 40, 20, 10, 5, 1, 0.5 are shown in Figure 5.4. The Airy half-wave length and half-wave period equivalents for a break width of 5 is 23 kilometers and about 1.8 minutes, a break width of 10 is 46 kilometers and 3.5 minutes, a break width of 20 is 90 kilometers and 7 minutes, and for a break width of 40 is 180 kilometers and 14 minutes.

The use of nonlinear shallow water models to describe tsunami waves generated from earthquake generated surface displacements is adequate for tsunamis generated by surface displacements that are at least ten times wider than the depth. The nonlinear shallow water wave becomes less realistic the further it travels from its source and the smaller the width-to-depth ratio. Non-linear shallow water models are adequate for large wave length and long period tsunamis such as generated by the 1964 Alaskan or the 1960 Chile earthquakes where the periods were about 30 minutes and wave length-to-depth ratio in the deep ocean was greater than 80. Tsunami waves generated by earthquakes with small areas and periods of a few minutes will not be realistically described using shallow water models. Either the linear gravity wave model or better the Navier-Stokes model should be used for accurate modeling of tsunamis with small (less than 10) width-to-depth ratios.

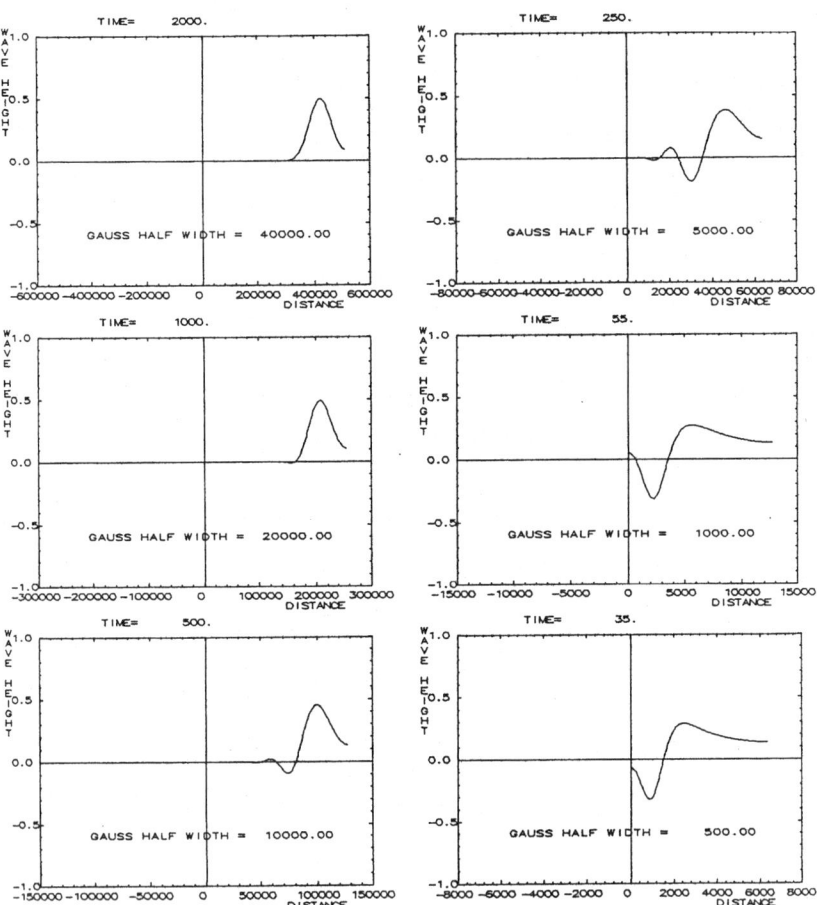

Fig. 5.4. The wave height in meters as a function of distance after the wave has traveled ten times its initial width for a 1 meter high Gaussian displacement with a Gaussian break length of 40, 20, 10, 5, 1, and 0.5 kilometers in 4450 meters deep water for the linear gravity wave model using the LGW code.

Sec. 5A TSUNAMI WAVE GENERATION

Generation Conclusions

The *SWAN* code solves the nonlinear, shallow water, long wave equations including the effects of friction and flooding. Both the *SWAN* and *ZUNI* code which solves the incompressible Navier-Stokes equations were used to study the development of a tsunami wave from an initial sea surface displacement. The development of a tsunami wave was also modeled using two-dimensional linear gravity wave theory.

If the initial displacement is approximately Gaussian and the wave length is very long compared to the depth, similar tsunami waves formed for all three methods. Two tsunami waves traveling in opposite directions formed that were the sum of the original surface displacement. However, dispersion effects resulted in Navier-Stokes and linear gravity waves with decreasing front slopes and amplitudes, and followed by a train of small waves.

The shallow water wave has a constant vertical velocity while the Navier-Stokes and linear gravity waves have the more realistic variable vertical velocity. For long wave length tsunamis the constant vertical velocity closely reproduces the velocity characteristics of Navier-Stokes and linear gravity waves, which slowly decrease with depth.

With decreasing periods and wave lengths, the discrepancy between the shallow water and the Navier-Stokes and linear gravity waves formed from initial sea surface displacement increases. These studies permit us to determine the parameteric region where the shallow water model can be useful. A summary of that region is presented at the end of this chapter.

Sources that involve compressible flow can not be solved with the incompressible models evaluated in this chapter; however they can be modeled using the compressible models described in Chapter 6.

Computer movies of the source calculations are in the NMWW CD-ROM TSUNAMI.MVE/SOURCE.MVE directory.

175

5B. Tsunami Wave Propagation

The shallow water *SWAN* code and the incompressible Navier-Stokes *ZUNI* codes were used to study the propagation of a tsunami wave from an initial sea surface displacement such as the propagation of a tsunami wave from Alaska or the U.S. West Coast to Hawaii. The propagation of a tsunami wave was also modeled using two-dimensional linear wave theory.

Tsunami Propagation Study – *SWAN* Model

A 1 meter high Airy half-wave surface displacement with a width of 45 kilometers in 4550 meters deep water was studied. This approximates within cell resolution a Gaussian wave with a Gaussian break width of 10 kilometers. Calculations were performed using a 250 meters wide mesh of 32,000 by 4 cells and at 0.5 sec intervals.

The wave height in meters as a function of distance is shown in Figure 5.5 at half hour intervals.

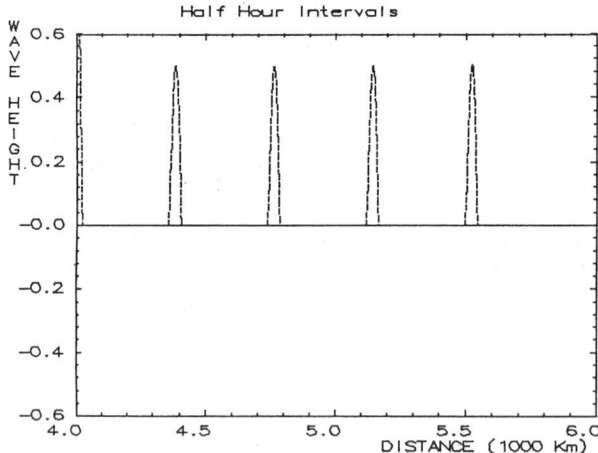

Fig. 5.5. The wave height in meters as a function of distance in meters at half hour intervals for a 1 meter high surface displacement with a half width of 45 kilometers in 4550 meters deep water for the nonlinear shallow water model using the *SWAN* code.

Sec. 5B TSUNAMI WAVE PROPAGATION

The initial surface displacement separates into two shallow water waves with a height of 0.5 meter and width of 45 kilometers. The nonlinear feature of the shallow water model included in the *SWAN* code results in a small trailing wave with an amplitude of less than 0.01 meter. The peak wave amplitude remains nearly constant as the wave propagates.

Tsunami Propagation Study – *ZUNI* Model

A 1 meter high approximately (within cell resolution) Gaussian surface water displacement with a Gaussian break width of 10 kilometers (which is equivalent to an Airy wave with a half-wave length of 45 kilometers) in 4550 meters deep water was studied as it traveled for 3 hours. The calculations for this geometry were performed with 15 cells in the Y or depth direction and up to 13,600 cells in the X or distance direction. The calculation required 231,200 cells and over 2.5 million mesh quantities.

The cells were rectangles 450 meters high in the Y direction and 4000 meters long in the X direction. The time increment was 3 sec. The water level was placed at 4550 meters or 50 meters up into the eleventh cell. The viscosity coefficient used was 0.02 poise, a value representative of the actual viscosity for water. The viscosity did not significantly affect the results.

The wave height in meters as a function of distance is shown in Figure 5.6 at various times. The initial surface displacement separates into two nearly shallow water waves with a height of 0.5 meter and width of 45 kilometers. At 100 sec the two waves have a vertical velocity at the top center of the wave of 0.0231 meter/sec. Near the bottom of the wave the velocity is 0.02016 meter/sec. An Airy wave with a 0.5 meter half-width, 90 kilometers wave length in 4550 meters of water has a vertical velocity of 0.0235 to 0.0226 meter/sec at the center of the wave. The Airy wave speed is 208.36 meters/sec and the group velocity is 202.89 meters/sec.

EVALUATION OF INCOMPRESSIBLE MODELS FOR MODELING WATER WAVES

Chap. 5

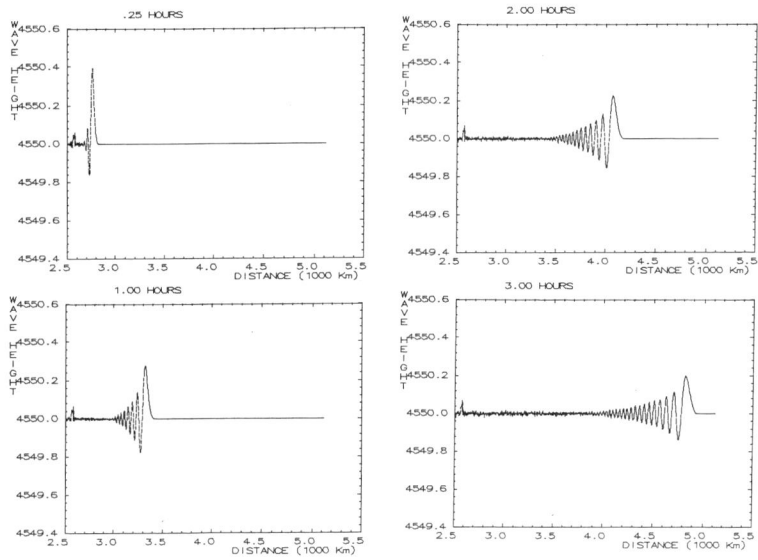

Fig. 5.6. The wave height in meters as a function of distance in kilometers at various times for a 1 meter high surface displacement with a width of 45 kilometers in 4550 meters deep water for the Navier-Stokes water wave model using the *ZUNI* code.

As the wave propagates, the wave height decreases, the slope of the front of the wave becomes smaller, and a train of small waves forms behind the main wave. After the wave has propagated for 3 hours and over 2.3 megameters, the wave height has decreased to 0.20 meters. A train of waves has developed behind the main wave with maximum negative amplitude of 0.14 meter and positive amplitude of 0.11 meter.

Tsunami Propagation Study – *LGW* Model

A 1 meter high Gaussian surface water displacement with a Gaussian break width of 10 kilometers (which is equivalent to an Airy wave with a half-wave length of 45 kilometers) propagating for 3 hours in 4550 meters deep water was studied.

Sec. 5B   TSUNAMI WAVE PROPAGATION

The wave height in meters as a function of distance is shown in Figure 5.7 at various times. The initial surface displacement separates into two nearly shallow water waves with a height of 0.5 meter and length of 45 kilometers.

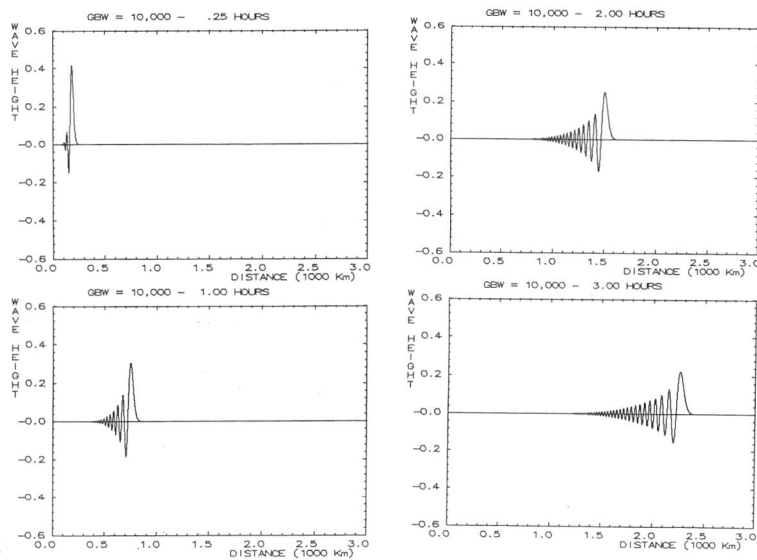

Fig. 5.7. The wave height in meters as a function of distance in kilometers at various times with a width of 45 kilometers in 4550 meters deep water for the linear gravity wave model using the LGW code.

The analytical linear gravity wave calculations were performed using up to 16,384 points. The wave profile was calculated for each time selected.

As the wave propagates, the wave height decreases, the slope of the front of the wave becomes smaller, and a train of small waves forms behind the main wave. After the wave has propagated for 3 hours and over 2.3 megameters, the wave height has decreased to 0.22 meter. A train of waves has developed behind the main wave with a maximum negative amplitude of 0.16 meter and positive amplitude of 0.12 meter. The peak height of the linear gravity wave decreases slower and the following wave train heights are higher than for the Navier-Stokes wave.

EVALUATION OF INCOMPRESSIBLE   Chap. 5
MODELS FOR MODELING WATER WAVES

The peak height of the Navier-Stokes wave is about 10 percent less than the linear gravity wave peak height after three hours of travel. This difference is an order of magnitude less than the difference between either wave and the shallow water wave height, which is more than twice as high. The period of the waves in the following wave train are similar; however the amplitudes are slightly higher for the linear gravity wave train. These results support the use of the linear gravity wave model for studying the effect of various wave length-to-depth ratios and of tsunami propagation for longer distances and times.

The linear gravity wave model was used to study wave propagation from initial surface displacements of width-to-depth ratios of 40 to 5.0 (Gaussian break widths of 40 to 5 kilometers) in 4550 meters deep water. The wave height after the wave had traveled for 0.5, 2.0, 5.0 and 10.0 hours for depth ratios of 40, 20, 10, and 5 are shown in Figure 5.8.

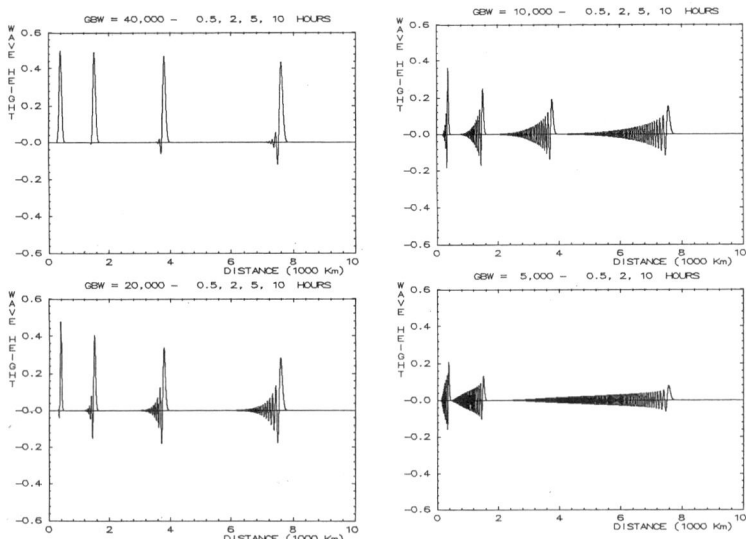

Fig. 5.8. The wave height in meters as a function of distance in meters after the wave has traveled up to 10 hours and 7.6 megameters for a 1 meter high Gaussian displacement with a Gaussian break length of 40, 20, 10, and 5 kilometers in 4550 meters deep water using the LGW code.

Sec. 5B   TSUNAMI WAVE PROPAGATION

The Airy half-wave length and half-wave period equivalents for a break width of 5 is 23 kilometers and about 1.8 minutes, a break width of 10 is 46 kilometers and 3.5 minutes, a break width of 20 is 90 kilometers and 7 minutes, and for a break width of 40 is 180 kilometers and 14 minutes. After traveling for 10 hours and 7.6 megameters, the peak wave amplitude for the 40 break width wave decreased from 0.5 to 0.42 meter, the 20 break width wave amplitude decreased to 0.28 meter, the 10 break width wave amplitude decreased to 0.16 meter, and the 5 break width wave amplitude decreased to 0.08 meter. The addition of nonlinear effects will lower these heights so these are upper bounds on the peak wave amplitude.

Propagation Conclusions

The nonlinear shallow water wave becomes less realistic the further it travels from its source and the smaller the width-to-depth ratio. Non-linear shallow water models are adequate for large wave length and long period tsunamis such as generated by the 1964 Alaskan or the 1960 Chile earthquakes where the periods were about 30 minutes and wave length to depth ratio in the deep ocean was greater than 80. Tsunami wave propagation from and generation by earthquakes with small areas and periods of a few minutes are not realistically described using shallow water models. Either the linear gravity wave model or better the Navier-Stokes model should be used for accurate modeling of long distance propagation of tsunamis with small (less than 40) width to depth ratios.

Most tsunami waves that have been observed after traveling across the ocean have periods longer than 10 minutes. These studies support the postulate that this is because the shorter wave length tsunamis are so dispersive that as they propagate long distances, their amplitude decreases by an order of magnitude.

EVALUATION OF INCOMPRESSIBLE Chap. 5
MODELS FOR MODELING WATER WAVES

Since three-dimensional Navier-Stokes solutions to tsunami propagation problems currently have limited practical application, most tsunami flooding studies need to be performed using the shallow water model. These studies permit us to determine the parameteric region where the shallow water model can be useful. A summary of that region is presented at the end of this chapter.

The development and application of three-dimensional Navier-Stokes models including a realistic bottom friction treatment will be required for significant improvement in our ability to realistically model tsunami generation, propagation and flooding. The current effort to develop such a capability is described in Chapter 6.

Computer movies of the propagation calculations are in the NMWW CD-ROM TSUNAMI.MVE/PROP.MVE directory.

5C. Tsunami Wave Flooding

The shallow water *SWAN* code and the Navier-Stokes *ZUNI* codes were used to study the effect of tsunami wave period, amplitude, bottom slope angle and friction on tsunami shoaling and flooding.

The shallow water waves shoal higher, steeper and faster than the Navier-Stokes waves. The differences increase as the periods become shorter and the slopes less steep with large differences for periods less than 500 sec and slopes less than 2 percent.

Tsunami Flooding Study – *SWAN* Model

A 3 meters half-height tsunami wave of various periods was propagated in 12 meters deep water and then interacted with slopes of various steepness.

The wave height in meters as a function of distance is shown in Figure 5.9 at various times for a tsunami Airy wave with a 900 sec period and a 3 meters half-height traveling 3000 meters in 12 meters deep water before it interacted with a frictionless 1 percent slope. The space resolution in the numerical model grid was 10 meters with 300 cells to the bottom edge of the slope, 120 cells on the slope below and 80 cells above the water surface. The calculations were performed at 0.5 sec intervals.

The peak wave height was 6.7 meters, the runup wave height was 6.0 meters and the inundation limit was 600 meters. The steepness of the slope was changed by changing the space resolution and the time step. The wave period was changed assuming that the tsunami was a shallow water Airy wave. The results are shown in Table 5.2 and Figures 5.10–5.12.

EVALUATION OF INCOMPRESSIBLE
MODELS FOR MODELING WATER WAVES

Chap. 5

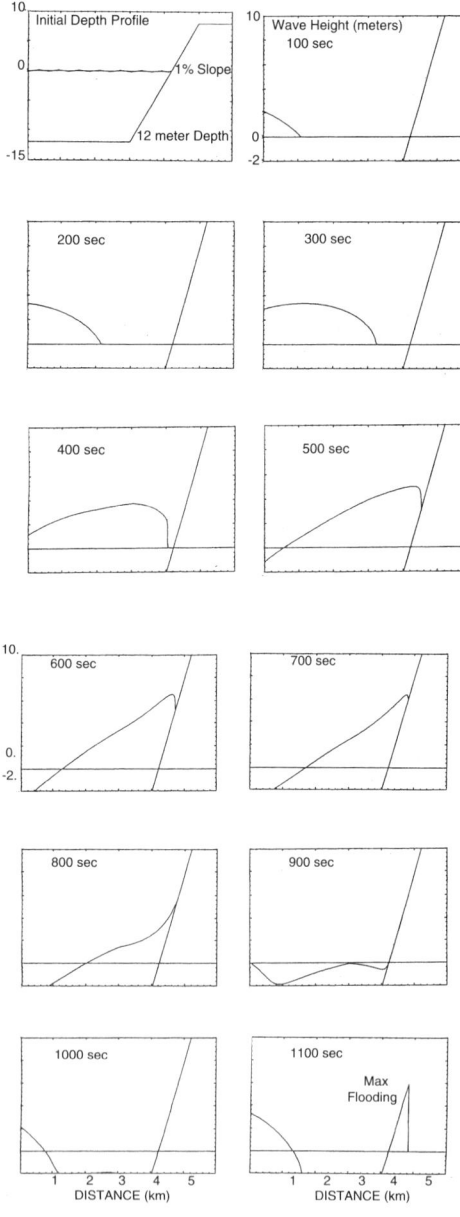

Fig. 5.9. The interaction of a 900 second period, 3 meters half-height tsunami in 12 meters of water with a 1 percent slope at various times. The last figure shows the first wave maximum extent of flooding.

TABLE 5.2
SWAN Shallow Water Flooding Study

Slope percent	Period seconds	DeChezy coef	Peak Ht. meters	Runup Ht. meters	Flood meters
	Initially 3 m Wave at 12 m Depth					
4.0	900.	0.0	6.8	6.3	160
2.0	900.	0.0	7.0	6.4	330
1.0	900.	0.0	6.7	6.0	600
0.5	900.	0.0	5.0	4.7	1000
0.25	900.	0.0	3.7	3.2	1300
1.0	1570.	0.0	7.2	6.3	600
1.0	450.	0.0	5.2	4.8	500
1.0	225.	0.0	3.6	3.2	350
2.0	1800.	0.0	6.0	5.6	280
2.0	450.	0.0	6.8	6.0	300
2.0	225.	0.0	5.0	4.8	250
4.0	450.	0.0	7.0	6.4	160
4.0	225.	0.0	6.6	6.0	160
2.0	900.	10.0	4.8	0.5	30
2.0	900.	20.0	5.8	1.8	100
2.0	900.	30.0	6.4	3.0	160
2.0	900.	50.0	6.6	5.2	280
1.0	900.	10.0	4.2	0.4	40
1.0	900.	15.0	5.0	0.8	80
1.0	900.	25.0	6.0	1.6	160
1.0	900.	40.0	6.5	2.8	300
1.0	900.	50.0	6.5	3.8	390
1.0	900.	60.0	6.5	4.4	440
100.	3600.	0.0	4.8	4.8	
100.	900.	0.0	6.1	6.1	
100.	225.	0.0	6.5	6.5	
	Initially 1 m Wave at 4550 m Depth					
5.0	2666.	0.0		3.5	
5.0	1333.	0.0		4.5	
5.0	999.	0.0		5.5	
5.0	666.	0.0		6.5	
5.0	330.	0.0		7.7	

EVALUATION OF INCOMPRESSIBLE Chap. 5
MODELS FOR MODELING WATER WAVES

The numerical model grid described is satisfactory for the cases studied giving runup and inundation values that are independent of space, time and geometry resolution. The grid model becomes less accurate for steeper slopes and longer wave length waves. The longer wave length waves need longer distances and times for the shoaling to occur; otherwise the calculated runup and inundation will be grid size dependent, decreasing with decreasing distance to the bottom edge of the slope.

The effect of the steepness of a frictionless slope on the runup height and inundation limit is shown in Figure 5.10 for a tsunami wave with a 3 meters half-height and a 900 sec period in 12 meters of water.

As shown in Figure 5.9, the highest shoaling wave height does not necessarily occur at the front of the wave. The difference between the height at the front of the wave and the peak wave height increases as the slope becomes steeper, with the maximum difference occuring for slopes larger than 1.5 percent. As shown in Figure 5.10, the runup height decreases and the inundation limit increases with decreasing slope angle.

Also shown in Figure 5.10 is the predicted runup height for the Bretschneider model without friction. Bretschneider's model results in smaller runup than $SWAN$ for slopes greater than 1 percent. A popular engineering method for estimating maximum probable tsunami inundation zones is to use historical inputs to provide tsunami wave heights as a function of frequency of occurrence and then to use the Bretschneider[6] model to calculate the runup on the shore at selected sites. The Bretschneider model as it is normally used includes surface roughness. The roughness parameters have been calibrated to reproduce observed runups. The Bretschneider model is independent of wave period and the only slope effect is from the friction.

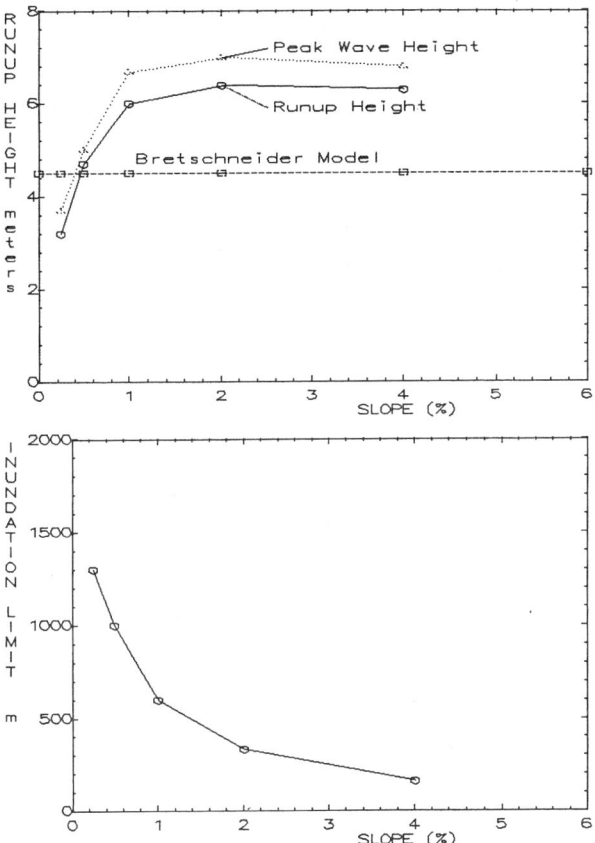

Fig. 5.10. The runup wave height and the inundation limit in meters as a function of slope for a 3 meters half-height 900 sec period tsunami wave in 12 meters of water for the nonlinear shallow water model using the *SWAN* code.

The effect of wave period and frictionless slope steepness on tsunami flooding is shown in Figure 5.11.

The maximum runup height is obtained for short period waves interacting with steep slopes, while the maximum inundation occurs for 1000 sec waves on gentler slopes.

EVALUATION OF INCOMPRESSIBLE   Chap. 5
MODELS FOR MODELING WATER WAVES

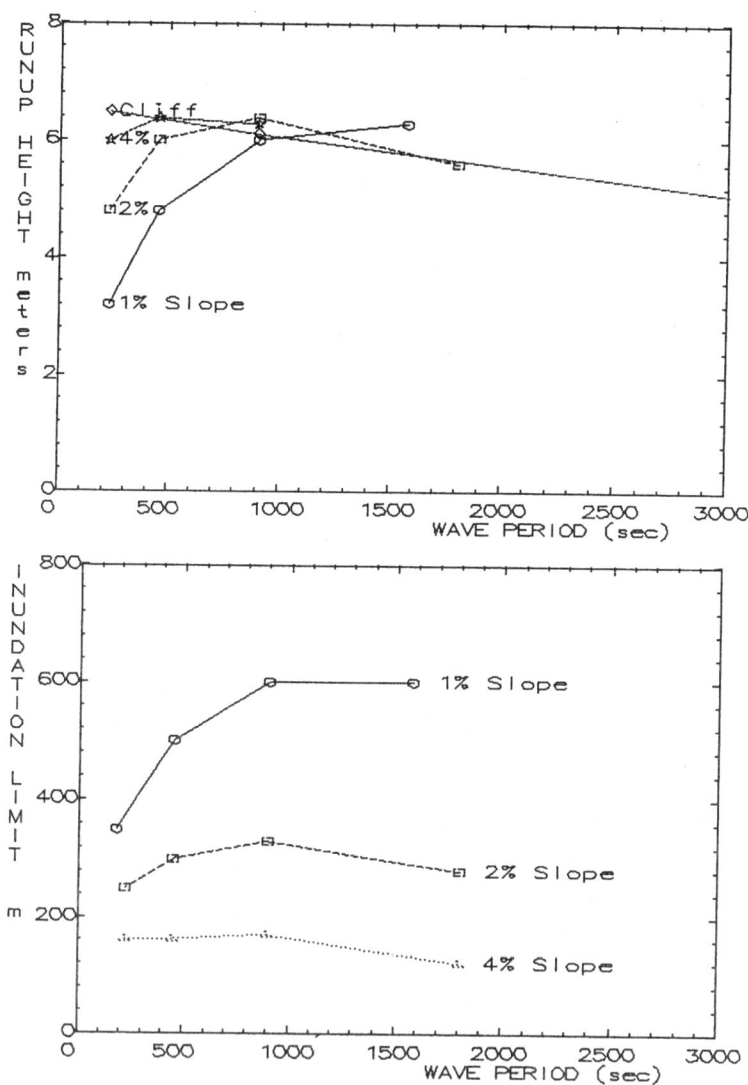

Fig. 5.11. The runup wave height and the inundation limit in meters as a function of wave period in sec for a 3 meters half-height 900 sec period tsunami wave in 12 meters of water for the nonlinear shallow water model using the *SWAN* code.

Sec. 5C   TSUNAMI WAVE FLOODING

The effect of friction on tsunami flooding of 1 and 2 percent slopes is shown in Figure 5.12. The effect of increasing friction (decreasing DeChezy friction constant) is small on the peak height and becomes larger on the runup height as the slope decreases. In the Bretschneider model, the surface roughness is described using a Manning "n" where 0.04 corresponds to a roughness characteristic of grass or small rocks and 0.03 to a roughness of many trees, large boulders or high grass. The DeChezy friction coefficient is related to Manning "n" by the depth to the 1/6 power. While not directly comparable, for depths in this study a DeChezy friction coefficient of 50 results in about the same friction effect as a Manning "n" in the 0.03–0.04 range.

Tsunami Flooding Study – *ZUNI* Model

A 1 meter high tsunami wave of various periods was propagated in 4550 meters deep water and then it interacted with frictionless slopes of various steepness. These results are an extension of those described in Chapter 3.

The geometry of the flooding calculation is shown in Figure 5.13 for a 1/15 or 6.66 percent slope. Calculations for this geometry were performed with 15 cells in the Y or depth direction and 68 cells in the X or distance direction. The cells were rectangles 450 meters high in the Y direction and 6750 meters long in the X direction. The time increment was 3 sec. The water level was placed at 4550 meters or 50 meters up into the eleventh cell. The viscosity coefficient used was 0.02 poise, a value representative of the actual viscosity for water. The viscosity did not significantly affect the results.

The slope in a *ZUNI* calculation is determined by the diagonal through the cell as discussed in the numerical methods section. A 6.66 percent slope results in a cell aspect ratio of 1 to 15 and a 4.0 percent slope an aspect ratio of 1 to 25. Numerical errors increase with increasing aspect ratio to unacceptable levels for 1 to 50 and greater ratios.

EVALUATION OF INCOMPRESSIBLE MODELS FOR MODELING WATER WAVES

Chap. 5

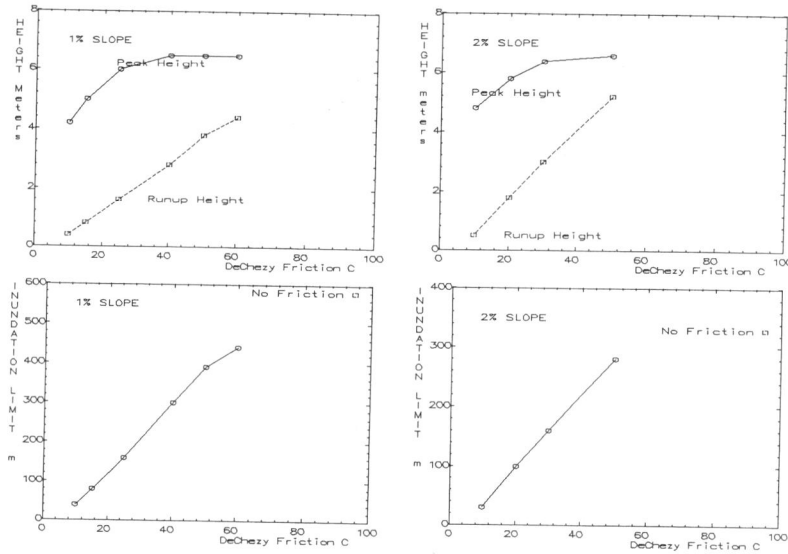

Fig. 5.12. The wave height and the inundation limit in meters for a 1 and 2 percent slope as a function of the DeChezy friction coefficient for a 3 meters half-height 900 sec period tsunami wave in 12 meters of water for the nonlinear shallow water model using the *SWAN* code.

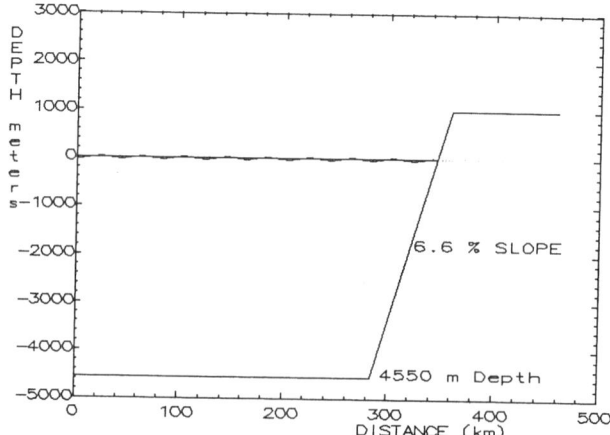

Fig. 5.13. The tsunami flooding geometry for the Navier-Stokes model using the *ZUNI* code.

TABLE 5.3
ZUNI Navier-Stokes Flooding Study

Slope	Period	First Peak	First Trough	2nd Peak	2nd Trough
percent	seconds	meters	meters	meters	meters

Period Study for 1 m Wave at 4550 m Depth

6.66	1333.	+3.2	−3.2	+4.1	−4.1
6.66	900.	+3.0	−4.0	+4.1	
6.66	500.	+1.9	−2.9	+2.9	−2.9
6.66	250.	+0.9	−0.5	+0.25	−0.1

Period Study for 1 m Wave at 1011 m Depth

1.48	2500.	+2.8	−3.3	+3.8	
1.48	1700.	+2.5	−3.7	+3.6	−3.2
1.48	1333.	+2.1	−3.6	+3.2	−3.3
1.48	900.	+1.5	−2.0	+2.0	
1.48	500.	+0.8	−0.3	+0.3	

Slope Study, Slope Width Varied, 4550 m Depth

10.0	1333.	+3.0	−3.2	+3.2	
6.6	1333.	+3.2	−3.9	+4.1	−4.1
4.0	1333.	+2.9	−4.0	+3.9	−3.8

Slope Study, Depth Varied, Slope Width Constant

13.48	1333.	+3.2	−3.3	+3.5	−3.0
6.66	1333.	+3.2	−3.9	+4.1	−4.1
2.96	1333.	+2.9	−3.9	+3.85	
2.22	1333.	+2.8	−3.8	+3.8	−3.7
1.48	1333.	+2.1	−3.6	+3.2	−3.3
0.74	1333.	+1.1	−1.9	+2.8	

EVALUATION OF INCOMPRESSIBLE Chap. 5
MODELS FOR MODELING WATER WAVES

TABLE 5.4

A 1333 Sec Tsunami Wave with Fixed Slope Width

Slope percent	Period seconds	Depth meters	Vel m/sec	Length km	Slope Width km
13.48	1333.	9100.	299.	398.	68.25
6.66	1333.	4550.	210.	280.	68.25
2.96	1333.	2022.	141.	188.	68.25
2.22	1333.	1517.	122.	163.	68.25
1.48	1333.	1011.	99.	132.	68.25
0.74	1333.	505.	70.	93.	68.25

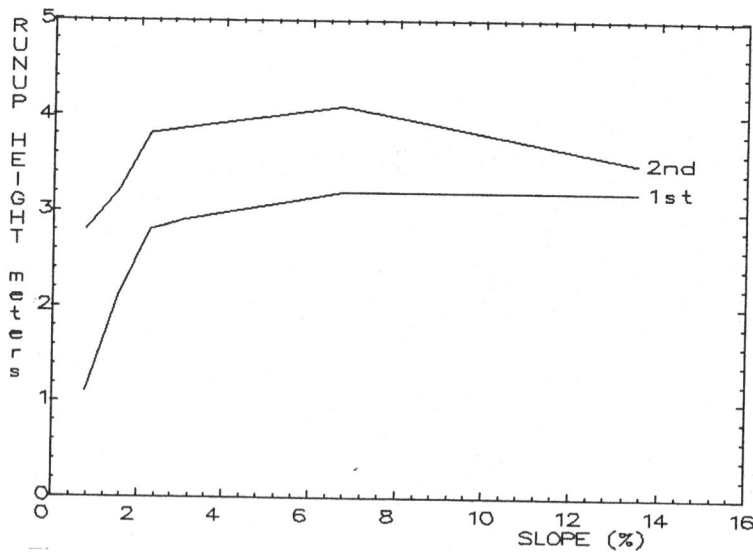

Fig. 5.14. The runup height as a function of slope for a 1 meter high, 1333 sec period tsunami for the Navier-Stokes model using the *ZUNI* code.

Sec. 5C   TSUNAMI WAVE FLOODING

Another method of changing the slope is to keep the width and aspect ratio constant and change the depth of the calculation. This permits slopes from 13.48 to 0.75 percent to be studied; however, if the period is kept constant, the other wave parameters change as shown in Table 5.4.

The slope study was performed using both methods to change the slope with the results shown in Table 5.3 and Figure 5.14. The general features of the calculated runup as a function of slope are the same for the *ZUNI* and *SWAN* models as shown in Figures 5.10 and 5.14. The resolution of the *ZUNI* calculation is inadequate to resolve the peak from the runup height. The *ZUNI* calculation permits us to realistically calculate wave interactions after the first wave interacts with the slope and to obtain rundown and runup for later waves. This is important because the largest runup both observed and in these calculations is the second or third wave. The interaction of the reflected waves with the later waves often results in the second or third runup being much different than the first runup. The magnitude and direction of the effect depends upon both the slope and the wave period as shown in Table 5.3. To perform these calculations, the source of the wave must be far removed from the slope in order to correctly follow the interaction of the incoming and reflected waves for several waves. A longer wave period needs a larger distance to run and requires more time for the multiple wave interactions. This results in large computing meshes and long running times for the long period tsunamis. Since they also result in the greatest flooding, they are also of the most interest.

The effect of wave period was investigated for a 1 meter half-height wave at 4550 meters interacting with a 6.66 percent slope and at 1011 meters depth with a 1.48 percent slope. The results of the study of wave period and slope are shown in Table 5.3 and Figure 5.15. Also shown in Figure 5.15 are the *SWAN* runup heights as a function of period for the first wave. To obtain adequate resolution in the *SWAN* calculation, a much smaller mesh is required than in the *ZUNI* calculation.

193

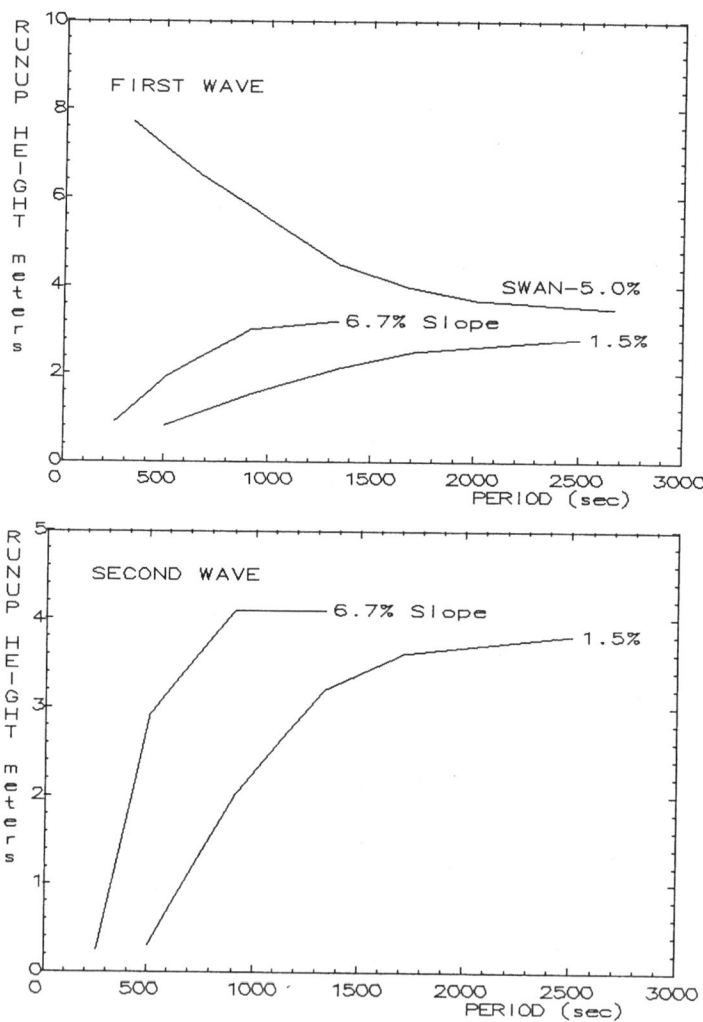

Fig. 5.15. The runup height as a function of period and different slopes and the first and second waves for a 1 meter high, 1333 sec period tsunami for the Navier-Stokes model using the *ZUNI* code. Also shown is the first wave using the *SWAN* shallow water code.

Sec. 5C   TSUNAMI WAVE FLOODING

This is a result of the particle surface following treatment in the *ZUNI* calculation which resolves the surface in each cell. The *SWAN* calculations were performed using a 780 cell long mesh, a mesh width of 500 meters and a time step of 0.3 sec. The 4500 meters deep bottom was 250 kilometers long (500 cells) to the edge of the 5 percent shelf formed by 90 kilometers (180 cells) below and 50 kilometers (100 cells) above the water surface. Calculations were also performed using a mesh width of 250 meters and a 1560 cell long mesh.

As shown in Table 5.3 and Figure 5.15, the shallow water waves flood higher for the first wave than the Navier-Stokes waves. The differences increase with shorter periods and less steep slopes, with large differences for periods less than 500 sec and slopes less than 2 percent. For higher period waves, the second wave is as much as a third larger than the first wave. The second wave is also larger than the shallow water first wave for wave periods greater than 500 sec.

Conclusions

The effect of tsunami wave period, amplitude, bottom slope angle and friction on tsunami shoaling and flooding has been investigated using a shallow water and a Navier-Stokes model.

The study shows higher wave shoaling and flooding for waves interacting with steeper slopes, for waves with longer periods and for waves from deeper water. The shallow water waves shoal higher, steeper and faster than the Navier-Stokes waves. The differences increase as the periods become shorter and slopes less steep with large differences for periods less than 500 sec and slopes less than 2 percent.

The interaction of the reflected first wave with the later waves often results in the second or third runup being much different than the first runup. The magnitude and direction of the effect depends upon both the slope and the wave period. For higher period waves, the second wave is as much as a third larger than the first wave.

EVALUATION OF INCOMPRESSIBLE MODELS FOR MODELING WATER WAVES
Chap. 5

Since the shallow water model distorts the shoaled and reflected wave, the wave interaction and resulting second and third runups are inaccurately modeled.

The effect of friction was investigated only for the shallow water model. It is the largest unknown factor in evaluating tsunami flooding. The Navier-Stokes model used does not have a treatment for friction.

The development and evaluation of a realistic friction model is an important remaining tsunami flooding problem. A possible solution to the problem is described in Chapter 6.

Since three-dimensional Navier-Stokes solutions to tsunami flooding problems currently have limited practical application, many tsunami flooding studies will need to be performed using the shallow water model. The following is a summary of the wave parameters where the shallow water model is valid.

SOURCE – wave length must be 10 times longer than the depth.

PROPAGATION – wave length must be 40 times longer than deep ocean depth and period needs to be longer than 15 minutes.

RUNUP – period needs to be longer than 15 minutes.

FLOODING – period needs to be longer than 10 minutes and the slope needs to be greater than 2 percent.

Many important water wave problems can not be adequately modeled using the shallow water model.

The incompressible Navier-Stokes model is adequate for many water wave problems that can not be accurately modeled using shallow water models. Currently only an upper limit flooding determination can be obtained without a realistic friction model. As shown in Chapter 6, it is now possible to model with high resolution the interface between the ocean and the sea floor, the mixing and movement of sediment and to include detailed sea floor topography. Future tsunami flooding studies can include friction effects realistically using the modeling methodology described in Chapter 6.

One of the more important water wave problems is the determination of Civil Defense evacuation zones for tsunamis. As shown in this chapter the shallow water model is inadequate for the task.

The incompressible Navier-Stokes model is unable to describe the initial generation of waves where the fluid-dynamics is compressible such as large earthquakes, impact landslides, volcanic, nuclear or conventional explosions and asteroids. Until recently there was no practical numerical method to solve such problems. Recent advances in the numerical modeling of water waves using the compressible Navier-Stokes model permit us to model almost any water wave generation, propagation and flooding problem. The next chapter describes the compressible model.

References

1. Charles L. Mader, "Numerical Tsunami Flooding Study – I," Science of Tsunami Hazards, Vol. 8, pp. 79–96 (1990).
2. Charles L. Mader, Dennis W. Moore, and George F. Carrier, "Numerical Tsunami Source Study – II," Science of Tsunami Hazards, Vol. 11, pp. 81–92 (1993).
3. Charles L. Mader, Dennis W. Moore, and George F. Carrier, "Numerical Tsunami Propagation Study – III," Science of Tsunami Hazards, Vol. 11, pp. 93–106 (1993).
4. G. F. Carrier and H. P. Greenspan, "Water Waves of Finite Amplitude on a Sloping Beach," Journal of Fluid Mechanics, Vol. 4, pp. 97–109 (1958).
5. G. F. Carrier, "The Dynamics of Tsunamis," Mathematical Problems in the Geophysical Sciences, American Mathematical Society, Vol. 1, pp. 157–187 (1971).
6. C. L. Bretschneider and P. G. Wybro, "Tsunami Inundation Prediction," Proceedings of 15th Coastal Engineering Conference (1976).

6

MODELING WAVES USING COMPRESSIBLE MODELS

Compressible fluid dynamics is modeled numerically using the Lagrangian and the Eulerian equations of motion. One, two and three-dimensional models and codes have been developed over the last fifty years using various numerical methods for solving the equations of motion and to describe the compressible equation of state.

In reference 1 is a complete description of Lagrangian and Eulerian approaches to modeling compressible reactive flows and the computer codes that were developed through the mid 1990's.

The Department of Energy's program Accelerated Strategic Computing Initiative (ASCI) of the last five years has resulted in major advances in computing technology and in methods for improving the numerical resolution of compressible hydrodynamic calculations. Only recently has it become possible to calculate water wave problems for minutes or even hours using compressible hydrodynamic models that require millions of time steps for each second of flow. Only recently has it become possible to finely resolve a water-air interface and follow the water wave with millimeter resolution in a problem with 40 kilometers of air, 5 kilometers of water and tens of kilometers of ocean crust. The limitations of incompressible flow discussed in the previous chapters no longer must be accepted for modeling water waves.

In this chapter the most advanced ASCI computer codes for modeling compressible fluid dynamics will be described. The codes will be used to model water waves generated from impact landslides, explosions, projectile impacts and asteroids. Numerical modeling of water waves has advanced so far so rapidly that it is clearly a technological revolution.

MODELING WAVES USING COMPRESSIBLE MODELS Chap. 6

6A. The Three-Dimensional Compressible Model

The three-dimensional partial differential equations for nonviscous, nonconducting, compressible fluid flow are

The Nomenclature

I	internal energy
P	pressure
U_x	velocity in x direction
U_y	velocity in y direction
U_z	velocity in z direction
ρ	density
t	time

$$\frac{\partial \rho}{\partial t} + U_x \left(\frac{\partial \rho}{\partial x}\right) + U_y \left(\frac{\partial \rho}{\partial y}\right) + U_z \left(\frac{\partial \rho}{\partial z}\right)$$

$$= -\rho \left(\frac{\partial U_x}{\partial x} + \frac{\partial U_y}{\partial y} + \frac{\partial U_z}{\partial z}\right),$$

$$\rho \left[\frac{\partial U_x}{\partial t} + U_x \left(\frac{\partial U_x}{\partial x}\right) + U_y \left(\frac{\partial U_x}{\partial y}\right) + U_z \left(\frac{\partial U_x}{\partial z}\right)\right] = -\frac{\partial P}{\partial x},$$

$$\rho \left[\frac{\partial U_y}{\partial t} + U_x \left(\frac{\partial U_y}{\partial x}\right) + U_y \left(\frac{\partial U_y}{\partial y}\right) + U_z \left(\frac{\partial U_y}{\partial z}\right)\right] = -\frac{\partial P}{\partial y},$$

$$\rho \left[\frac{\partial U_z}{\partial t} + U_x \left(\frac{\partial U_z}{\partial x}\right) + U_y \left(\frac{\partial U_z}{\partial y}\right) + U_z \left(\frac{\partial U_z}{\partial z}\right)\right] = -\frac{\partial P}{\partial z},$$

$$\rho \left[\frac{\partial I}{\partial t} + U_x \left(\frac{\partial I}{\partial x}\right) + U_y \left(\frac{\partial I}{\partial y}\right) + U_z \left(\frac{\partial I}{\partial z}\right)\right]$$

$$= -P \left(\frac{\partial U_x}{\partial x} + \frac{\partial U_y}{\partial y} + \frac{\partial U_z}{\partial z}\right).$$

Sec. 6A   THE THREE-DIMENSIONAL COMPRESSIBLE
                                                   MODEL

These equations are solved by a high-resolution Godunov differencing scheme using an adaptive grid technique described in reference 2. The solution technique uses Continuous Adaptive Mesh Refinement (CAMR). The decision to refine the grid is made cell-by-cell continuously throughout the calculation. The computing is concentrated on the regions of the problem which require high resolution.

Refinement occurs when gradients in physical properties (density, pressure, temperature, material constitution) exceed defined limits, down to a specified minimum cell size for each material. The mesh refinement is shown in Figure 6.1 for a complicated geometry embedded in a material.

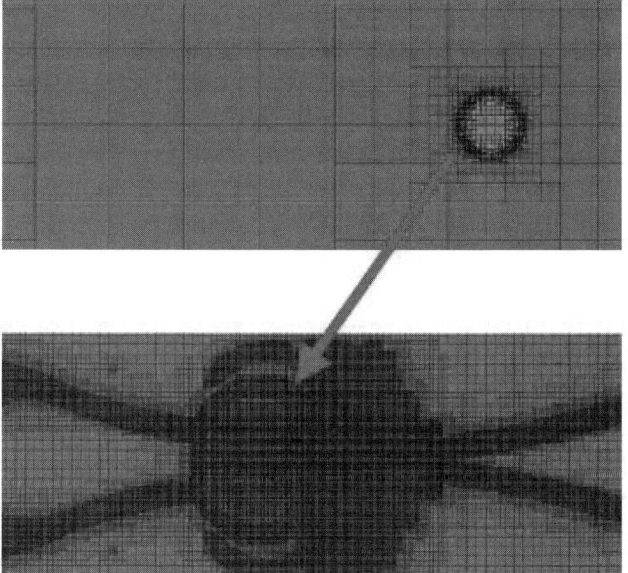

Fig. 6.1. Adaptive mesh refinement applied to a complicated geometry.

Much larger computational volumes, times and differences in scale can be simulated than possible using previous Eulerian techniques such as those described in reference 1.

The original code was called *SAGE*. A later version with radiation is called *RAGE*. A recent version with the techniques for modeling reactive flow described in reference 1 is called *NOBEL* and was used for modeling the Lituya Bay impact landslide generated tsunami and water cavity generation.

201

The codes can describe one-dimensional slab or spherical geometry, two-dimensional slab or cylindrical geometry, and three-dimensional Cartesian geometry.

Because modern supercomputing is currently done on clusters of machines containing many identical processors, the parallel implementation of the code is very important. For portability and scalability, the codes use the Message Passing Interface (MPI). Load leveling is accomplished through the use of an adaptive cell pointer list, in which newly created daughter cells are placed immediately after the mother cells. Cells are redistributed among processors at every time step, while keeping mothers and daughters together. If there are a total of M cell and N processors, this technique gives nearly M/N cells per processor. As neighbor cell variables are needed, the MPI gather/scatter routines copy those neighbor variables into local scratch memory.

The codes incorporate multiple material equations of state (analytical or SESAME tabular). Every cell can in principle contain a mixture of all the materials in a problem assuming that they are in pressure and temperature equilibrium. As described in reference 1, pressure and temperature equilibrium is appropriate only for materials mixed molecularly. The assumption of temperature equilibrium is inappropriate for mixed cells with interfaces between different materials. The errors increase with increasing density differences. While the mixture equations of state described in reference 1 would be more realistic, the problem is minimized by using fine numerical resolution at interfaces. The amount of mass in mixed cells is kept small resulting in small errors being introduced by the temperature equilibrium assumption. The strength is treated using an elastic-plastic model identical to that described in reference 1.

Very important for water cavity generation and collapse and the resulting water wave history is the capability to initialize gravity properly, which is included in the code. This results in the initial density and initial pressure changing going from the atmosphere at 40 kilometers altitude down to the ocean surface. Likewise the water density and pressure changes correctly with ocean depth.

Some of the remarkable advances in fluid physics using the *SAGE* code have been the modeling of Richtmyer-Meshkov and shock induced instabilities described in references 3 and 4.

6B. The Lituya Bay Mega-Tsunami

Lituya Bay, Alaska is on the northeast shore of the Gulf of Alaska. It is an ice-scoured tidal inlet with a maximum depth of 220 meters and a narrow entrance with a depth of only 10 meters. It is a T-shaped bay, 7 miles long and up to 2 miles wide. The two arms at the head of the bay, Gilbert and Crillon Inlets, are part of a trench along the Fairweather fault. On July 8, 1958, a 7.5 Magnitude earthquake occurred along the Fairweather fault with an epicenter near Lituya Bay.

A mega-tsunami wave was generated that washed out trees to a maximum altitude of 520 meters at the entrance of Gilbert Inlet. Much of the rest of the shoreline of the bay was denuded by the tsunami from 30 to 200 meters altitude as shown in Figures 6.2 – 6.4.

Fig. 6.2. The view of Lituya Bay in August 1958 after the earthquake and tsunami. The forest was destroyed to a maximun elevation of 524 meters and a maximum distance of 1000 meters from shoreline. The bay is 7 miles long and up to 2 miles wide. The two arms at the head of the bay are part of the Fairweather fault.

MODELING WAVES USING COMPRESSIBLE MODELS   Chap. 6

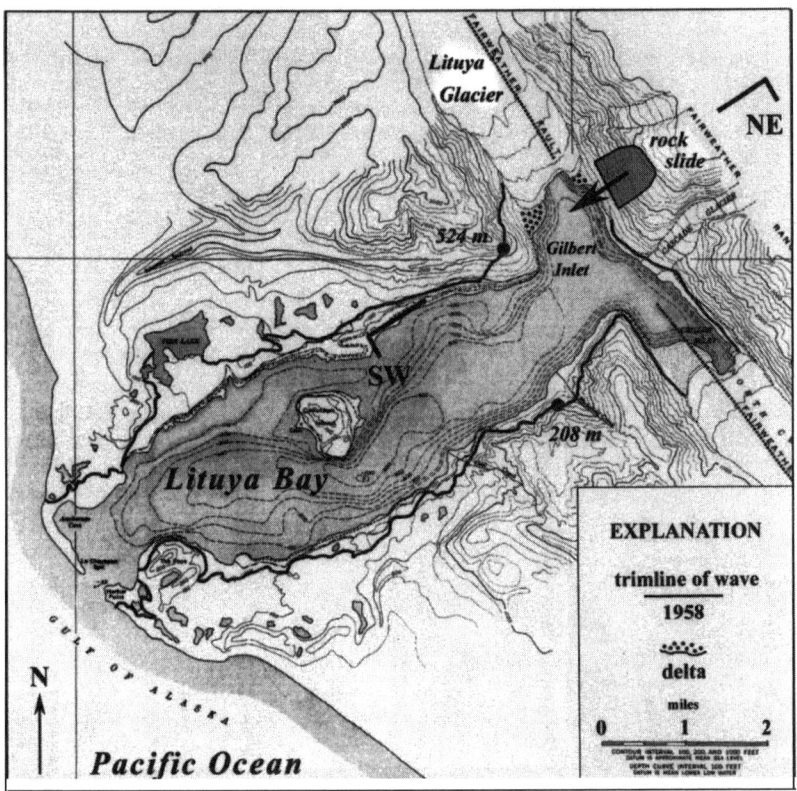

Fig. 6.3. Lituya Bay map showing topographic and bathymetric contours, trace of Fairweather fault, 1958 rockslide and trimline of wave runup. Forests destroyed to maximum elevations of 524 meters and 208 meters on north and south shores.

During the last 150 years five giant waves have occurred in Lituya Bay. The previous event occurred on October 27, 1936 which washed out trees to a maximum altitude of 150 meters and was not associated with an earthquake.

Don Miller recorded all that was known in 1960 about the giant waves in Lituya Bay in reference 5.

The July 9, 1958 earthquake occured at about 10:15 p.m., which is still daylight at Lituya Bay. The weather was clear and the tide was ebbing at about plus 5 feet.

Sec. 6B   THE LITUYA BAY MEGA-TSUNAMI

Bill and Vivian Swanson were on their boat anchored in Anchorage Cove near the western side of the entrance of Lituya Bay. Their astounding observations are recorded in reference 6 and were as follows:

"With the first jolt, I tumbled out of the bunk and looked toward the head of the bay where all the noise was coming from. The mountains were shaking something awful, with slides of rock and snow, but what I noticed mostly was the glacier, the north glacier, the one they call Lituya Glacier.

I know you can't ordinarily see that glacier from where I was anchored. People shake their heads when I tell them I saw it that night. I can't help it if they don't believe me. I know the glacier is hidden by the point when you're in Anchorage Cove, but I know what I saw that night, too.

The glacier had risen in the air and moved forward so it was in sight. It must have risen several hundred feet. I don't mean it was just hanging in the air. It seems to be solid, but it was jumping and shaking like crazy. Big chunks of ice were falling off the face of it and down into the water. That was six miles away and they still looked like big chunks. They came off the glacier like a big load of rocks spilling out of a dump truck. That went on for a little while — it's hard to tell just how long — and then suddenly the glacier dropped back out of sight and there was a big wall of water going over the point. The wave started for us right after that and I was too busy to tell what else was happening up there."

A 15 meters high wave rushed out of the head of the bay toward the Swansons' anchored boat. The boat shot upward on the crest of the wave and over the tops of standing spruce trees on the entrance spit of Lituya Bay. Bill Swanson looked down on the trees growing on the spit and said he was more than 25 meters above their tops. The wave crest broke just outside the spit and the boat hit bottom and foundered some distance from the shore. Swanson saw water pouring over the spit, carrying logs and other debris. The Swansons escaped in their skiff to be picked up by another fishing boat 2 hours later.

MODELING WAVES USING COMPRESSIBLE MODELS Chap. 6

The front of Lituya Glacier on July 10 was a nearly straight, vertical wall almost normal to the trend of the valley. Comparisions with photographs of the glacier taken July 7 indicate that 400 meters of ice had been sheared off of the glacier front. The photograph taken after the event is shown in Figure 6.4.

Fig. 6.4. The view of Lituya Bay in August 1958 where the forest was destroyed to a maximun elevation of 524 meters.

After the earthquake there was a fresh scar on the northeast wall of Gilbert Inlet, marking the recent position of a large mass of rock that had plunged down the steep slope into the water. The next day after the earthquake and tsunami, loose rock debris on the fresh scar was still moving at some places, and small masses of rock still were falling from the rock cliffs near the head of the scar.

The wave destroyed the forests and even removed the barnacles from the rocks. When the author visited Lituya Bay forty years after the event, almost no vegetation grew on either the slide region or on the cleared area shown in the above photograph.

Sec. 6B   THE LITUYA BAY MEGA-TSUNAMI

The dimensions of the slide on the slope are accurate but the thickness of the slide mass normal to the slope can only be estimated. The main mass of the slide was a prism of rock that was 730 meters and 900 meters along the slope with a maximum thickness of 90 meters and average thickness of 45 meters normal to the slope. The center of gravity was at about 600 meters altitude. As described in reference 5 this results in an approximate volume of 30 million cubic meters (40 million cubic yards).

Miller[5] concluded that "the rockslide was the major, if not the sole cause of the 1958 giant wave." The Swanson observations have not been believed as they indicate that a lot more than a simple landslide occurred.

Shallow Water Modeling

In reference 7 shallow water modeling was performed using the *SWAN* nonlinear shallow water code. The generation and propagation of the tsunami wave of July 8, 1958 in Lituya Bay was modeled using a 92.75 by 92.75 meters grid of the topography. The 3 by 6 second land topography was generated from the Rocky Mountain Communication's CD-ROM compilation of the Defense Mapping Agency (DMA) 1 by 1 degree blocks of 3 arc second elevation data. The sea floor topography was taken from sea floor topographic maps published in reference 5. The grid was 150 by 150 cells and the time step was 0.15 sec.

It was concluded that displaced water by a simple landslide or an earthquake along the Fairweather fault at the head of the bay could not result in the observed 550 meters high runup.

The water in the inlet with the width of the landslide and between the landslide and the 520 meter high runup was sufficient to cover the runup region to a 100 meters height. In reference 7 it was shown that this high water layer was sufficient to form a wave that will reproduce the observed flooding of the bay beyond the inlet.

It was concluded that a landslide impact model was required similar to that for asteroid generated waves. In 1999 it appeared that the numerical technology required would not become available for many decades.

MODELING WAVES USING COMPRESSIBLE MODELS

Physical Modeling

During the Summer of 2000, Hermann Fritz[8] conducted experiments that reproduced the 1958 Lituya Bay event. The 1:675 scale laboratory model of Lituya Bay shown in Figure 6.5 was built at VAW at the Swiss Federal Institute of Technology at Zurich, Switzerland. The laboratory experiments indicated that the 1958 Lituya Bay 524 meters runup on the spur ridge of Gilbert Inlet could have been caused by a landslide impact. The study was reported in reference 8. A novel pneumatic landslide generator was used to generate a high-speed granular slide with a controlled impact velocity and shape. A granular slide with the density and volume given by Miller[5] was impacted with a mean velocity of 110 meters/sec. It generated a large air cavity and an extremely nonlinear wave with a maximum scaled height of about 160 meters which ran up to a scaled elevation of 530 meters above mean sea level.

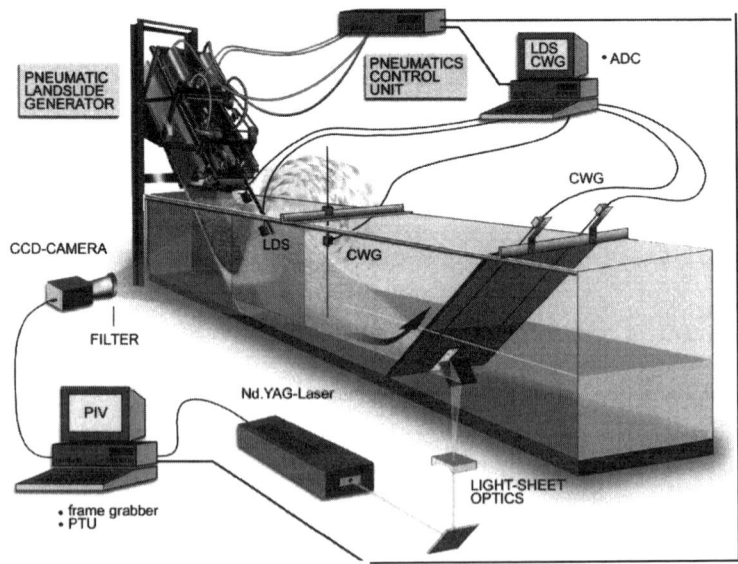

Fig. 6.5. Experimental setup with pneumatic installation and measurement systems that included laser distance sensors (LDS), capacitance wave gauges (CWG) and particle image velocimetry (PIV).

The initial geometry used by Fritz[8] in his experimental modeling in reference 8 is shown in Figure 6.6.

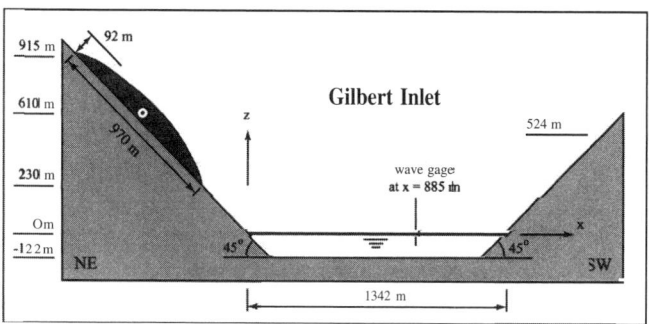

Fig. 6.6. Cross section of Gibert Inlet along slide axis used in the physical model.

The rockslide was simulated using 4 mm diameter Barium Sulfate grains with a density of 1.6 g/cc and a void fraction of 39 percent. To what extent the rockslide broke up before impacting Gilbert Inlet is unknown. Scanned slide impact profiles were used to calculate the mean slide impact velocity of 110 meters/sec. The slide profile showed a gentle increase in slide thickness with time to scaled maximum of 134 meters and a fast decay back to zero. The scaled maximum slide thickness of 134 meters is 1.4 times the pre-motion slide thickness of 92 meters estimated roughly by Miller in reference 5. The three phases – granular material, water and air – are clearly separated along distinct border lines before flow reattachment occurs. Flow reattachment traps a large volume of air in the back of the rockslide, which leads to large bubble formation, bubble breakup and massive mixing. The slide is deformed due to impact and deflection at the channel bottom.

A sequence of twelve instantaneous velocity vector fields computed the PIV data are shown in Figure 6.7. The sequence starts at 0.76 sec after rockslide impact and continues with a time step of 1.73 sec.

MODELING WAVES USING COMPRESSIBLE MODELS
Chap. 6

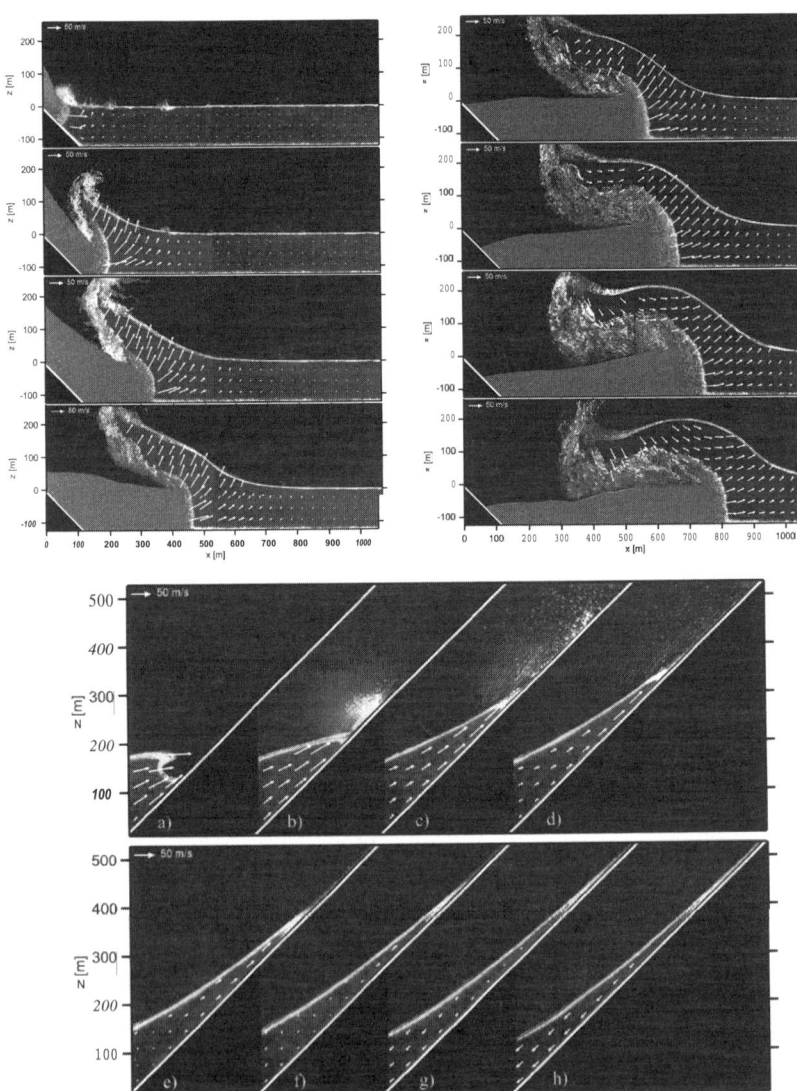

Fig. 6.7. Photosequence of the Fritz granular impact slide experiment. The time interval is 1.73 sec with the first image at 0.76 sec after impact.

Sec. 6B   THE LITUYA BAY MEGA-TSUNAMI

Compressible Navier-Stokes Modeling

The Lituya Bay impact landslide generated tsunami was modeled in reference 9 with the Navier-Stokes AMR (Adaptive Mesh Refinement) Eulerian compressible hydrodynamic code *NOBEL* including the effects of gravity. The geometry of the calculation is shown in Figure 6.8 superimposed on the actual topography.

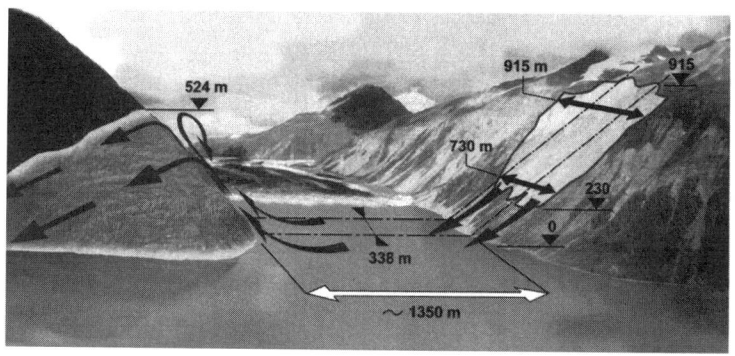

Fig. 6.8. Gilbert Inlet illustration showing rockslide dimensions, impact site and wave runup to 524 meters on the spur ridge directly opposite to the rockslide impact. Direction of view is North. The front of the Lituya Glacier is set to the 1958 post slide position. The preslide position extended 400 meters closer to the entrance of Gilbert Inlet.

The calculated density profiles are shown in Figures 6.9 and 6.10 for a rockslide moving with a resultant velocity of 110 meters/sec (X and Y component velocities of 77.8 meters/sec). The rockslide had an area of 21,000 square meters and was basalt with a density of 2.868 g/cc. The initial slide profile was triangular with its apex near the back of the slide to reproduce the Fritz slide profile shape described previously. The initial water depth was 120 meters and the length was 1.4 kilometers. The calculated maximum wave height in the bay was about 250 meters above sea level which ran up to 580 meters which is to be compared to the observed 524 meters. Such a wave could lift the glacier ice and result in the spectacular behavior observed by Swanson.

MODELING WAVES USING COMPRESSIBLE MODELS

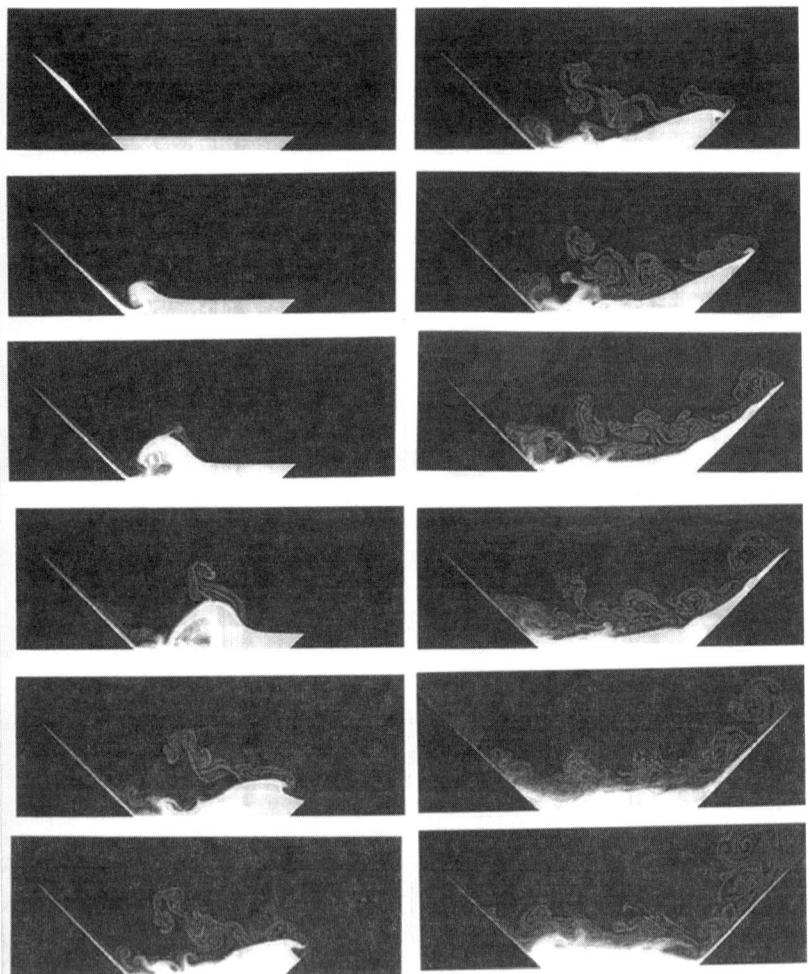

Fig. 6.9. Calculated density profiles of the Lituya impact landslide generated tsunami. The times are 0, 6, 10, 16, 22, 24, 26, 30, 36, 40, 48 and 58 sec.

A PowerPoint presentation with Fritz's experimental movies and computer animations is available on the NMWW CD-ROM in the NOBEL/LITUYA directory.

Sec. 6B THE LITUYA BAY MEGA-TSUNAMI

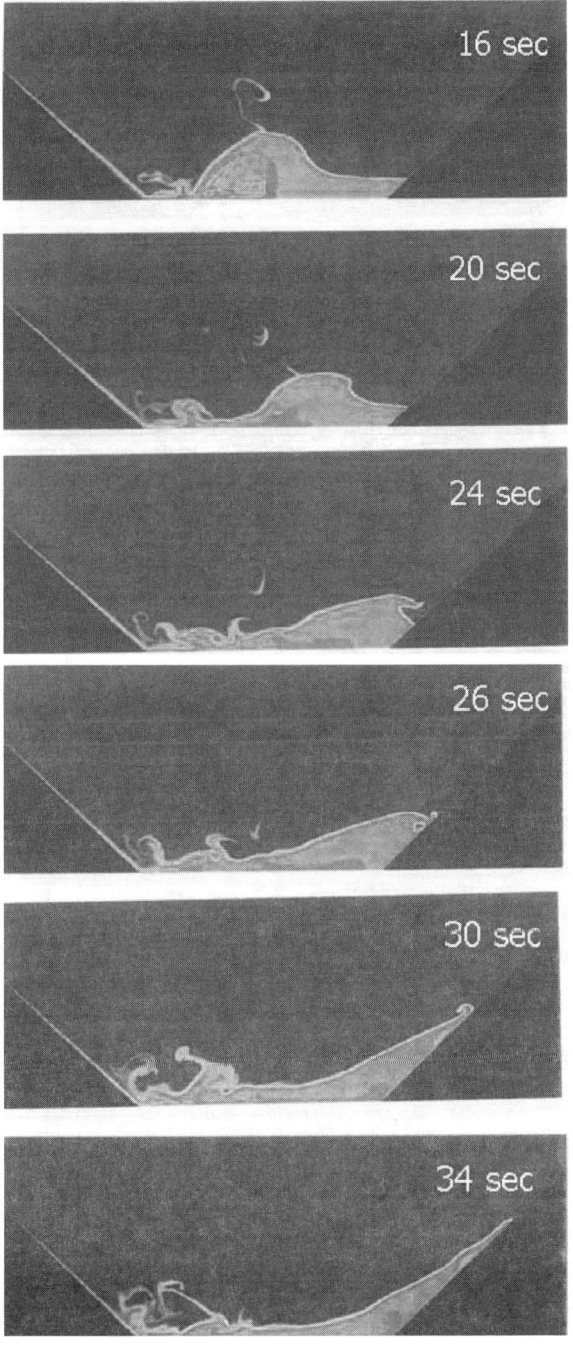

Fig. 6.10. continued on next page.

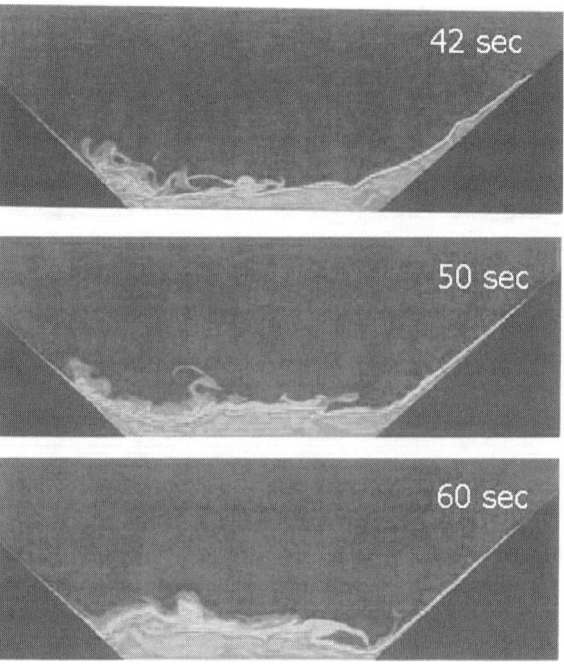

Fig. 6.10. Calculated density profiles of the Lituya impact landslide generated tsunami. The slide material is shown as a dark region running across the ocean floor. At later times it comes to rest mostly in the middle of the bay. This illustrates how friction effects can be modeled realistically.

Conclusions

The mega-tsunami that occurred on July 8, 1958 in Lituya Bay washed out trees to a maximum altitude of 520 meters at the entrance of Gilbert Inlet. Much of the rest of the shoreline of the bay was denuded by the tsunami from 30 to 200 meters altitude.

The Lituya Bay impact generated tsunami was modeled in reference 9 with the AMR Eulerian compressible hydrodynamic code called *NOBEL/SAGE*. The capability now exists to evaluate the potential impact landslide tsunami hazards for vunerable regions of the world.

6C. Water Cavity Generation

Introduction

In the mid 1960's, B. G. Craig[10] at the Los Alamos National Laboratory performed experiments designed to characterize the formation of water waves from explosives detonated near the water surface. He reported observing the formation of ejecta water jets above and jets or "roots" below the expanding gas cavity. This was unexpected and a scientific mystery which has remained unsolved until it was finally modeled using the *NOBEL* code in December of 2002.

In the early 1980's, the hypervelocity impact (1.25 to 6 kilometers/sec) of projectiles into water was studied at the University of Arizona by Gault and Sonett[11]. They observed quite different behavior of the water cavity as it expanded when the atmospheric pressure was reduced from one to a tenth atmosphere. Above about a third of an atmosphere, a jet of water formed above the expanding cavity and a jet or "root" emerged below the bottom of the cavity.

In the mid 1980's, similar results were observed by Kedrinskii[12] at the Institute of Hydrodynamics in Novosibirsk, Russia, who created cavities in water by sending large electrical currents through small lengths of small diameter Gold wires (bridge wires) causing the Gold to vaporize. The "exploding bridge wire" is a common method used to initiate propagating detonation in explosives. He observed water ejecta jets and roots forming for normal atmospheric pressure and not for reduced pressures.

Thus the earlier Craig observations were not caused by some unique feature of generation of the cavity by an explosion. The process of cavity generation by projectiles or explosives in the ocean surface and the resulting complicated fluid flows has been an important unsolved problem for over 50 years. The prediction of water waves generated by large-yield explosions and asteroid impacts has been based on extrapolation of empirical correlations of small-yield experimental data or numerical modeling assuming incompressible flow, which does not reproduce the above experimental observations.

The *NOBEL* code has been used to model the experimental geometries of Sonett and of Craig. The experimental observations were reproduced as the atmospheric pressure was varied as described in reference 13.

Projectile and Exploding Bridge Wire Generated Cavities

In the early 1980's, experiments were being performed at the University of Arizona to simulate asteroid impacts in the ocean. The hypervelocity impact (1.25 to 6 kilometers/sec) of various solid spherical projectiles (Pyrex or Aluminum) with water was performed by Gault and Sonett[11]. Their observations were similar to those previously observed by Craig[10]. While the water cavity was expanding, an ejecta jet was formed at the axis above the water plume and a jet or "root" emerged along the axis below the cavity. The water cavity appeared to close and descend into deeper water.

To improve the photographic resolution and reduce the light from the air shock, Sonett repeated his impact experiments under reduced atmospheric pressure. Much to everyone's surprise, when the pressure was reduced from one to a tenth atmosphere, the ejecta jet and the root did not occur and the water cavity expanded and collapsed upward toward the surface. This was what had been expected to occur in both the earlier Craig experiments and the projectile impacts.

It became evident that the atmospheric pressure and the pressure differences inside and outside the water plume above the water surface was the cause of the formation of the jet, the root, the cavity closure and descent into deeper water.

Figure 6.11 shows the Gault and Sonett[11] results for a 0.25 cm diameter aluminum projectile moving at 1.8 kilometers/sec impacting water at one atmosphere (760 mm), and at 16 mm air pressure. A water stem and jet occurs at one atmosphere and not at low pressure.

Figure 6.12 shows the Gault and Sonett results for a 0.635 cm diameter aluminum projectile moving at 2.5 kilometers/sec impacting water at one atmosphere (760 mm) and a 0.3175 cm diameter Pyrex projectile moving at 2.32 kilometers/sec impacting water at 16 mm air pressure. A water stem and jet occurs at one atmosphere and not at low pressure.

Sec. 6C   WATER CAVITY GENERATION

Professor Kedrinskii[12] at the Russian Institute for Hydrodynamics was also studying the generation of water cavities from exploding bridge wires. He was observing the formation of ejecta jets and roots as the water cavity expanded similar to those observed by Craig using explosives and by Gault and Sonett using projectiles. After we showed him the effect of reduced atmospheric pressure, he proceeded to repeat his exploding bridge wire experiments under reduced pressure. He observed that the jets and roots did not form when the atmospheric pressure was reduced to 0.2 atmosphere.

Figure 6.13 shows the Kedrinskii results for an exploding bridge wire in water at one atmosphere and at 0.2 atmosphere air pressure.

Compressible Navier-Stokes Modeling

The projectile impact and explosive generated water cavity was modeled with the recently developed full Navier-Stokes AMR (Adaptive Mesh Refinement) Eulerian compressible hydrodynamic code *NOBEL* described earlier. The continuous adaptive mesh refinement permits the following of shocks and contact discontinuities with a very fine grid while using a coarse grid in smooth flow regions. It can resolve the water plume and the pressure gradients across the water plume and follow the generation of the water ejecta jet and root.

Figure 6.14 shows the calculated density profiles for a 0.25 cm diameter aluminum projectile moving at 2.0 kilometers/sec impacting water at five atmosphere air pressure. The water plume collapses at the axis creating a jet moving upward and downward. The jet passes down through the cavity, penetrating the bottom of the cavity at the axis forming the stem. The flow results in the cavity descending down into the water.

Figure 6.15 shows the calculated density profiles for a 0.25 cm diameter aluminum projectile moving at 2.0 kilometers/sec impacting water at one atmosphere air pressure.

Figure 6.16 shows the calculated density profiles for a 0.25 cm diameter aluminum projectile moving at 2.0 kilometers/sec impacting water at 0.1 atmosphere air pressure. The tip of the water plume continues to expand in contrast to what is observed at atmospheric pressures higher than 200 mm.

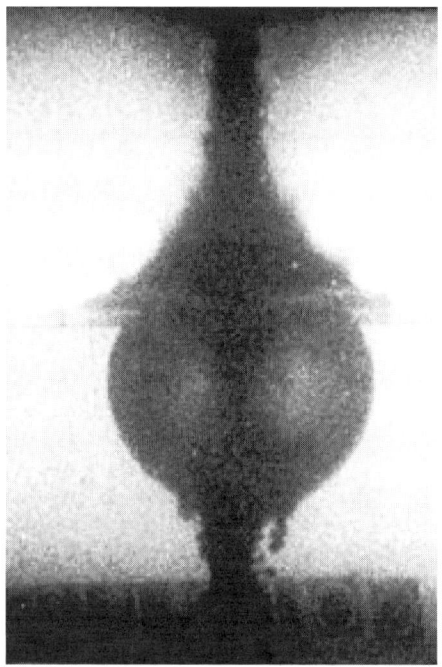

760 mm AIR PRESSURE

16 mm AIR PRESSURE

Fig. 6.11. The Gault and Sonett experimental results for a 0.635 cm diameter aluminum projectile moving at 1.8 kilometers/sec impacting water.

Sec. 6C   WATER CAVITY GENERATION

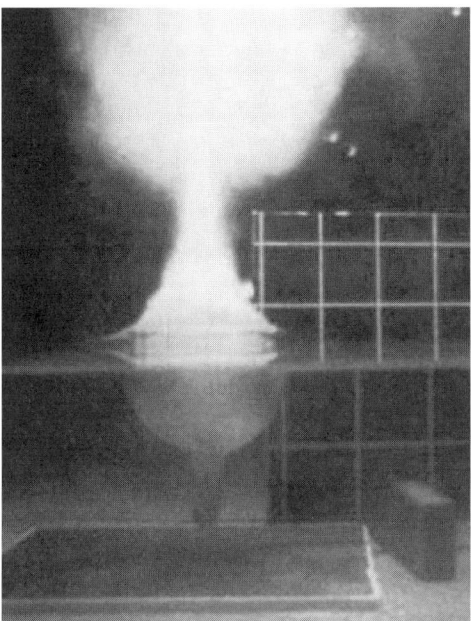

760 mm AIR PRESSURE

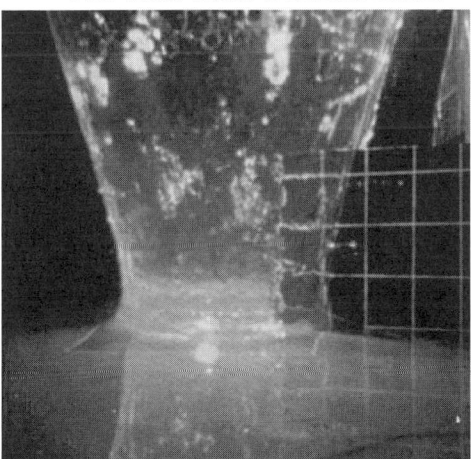

16 mm AIR PRESSURE

Fig. 6.12. The Gault and Sonett experimental results for a 0.25 cm diameter aluminum projectile moving at 2.5 kilometers/sec at 760 mm air pressure. A 0.3175 diameter Pyrex projectile moving at 2.32 kilometers/sec in 16 mm of air is shown in the bottom frame.

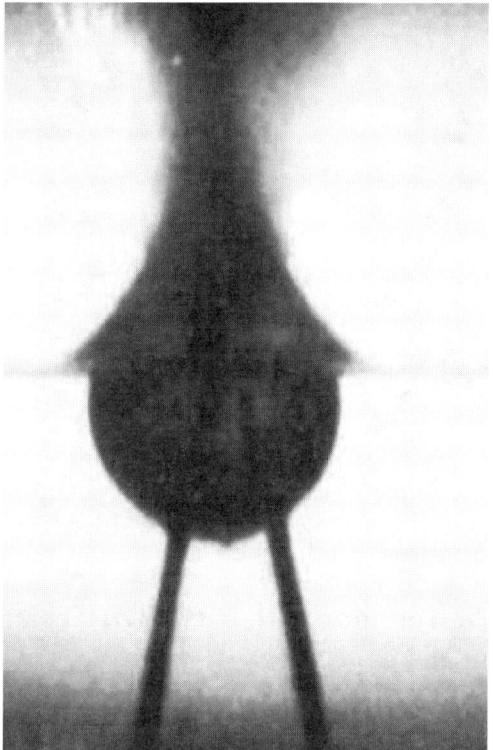

760 mm AIR PRESSURE

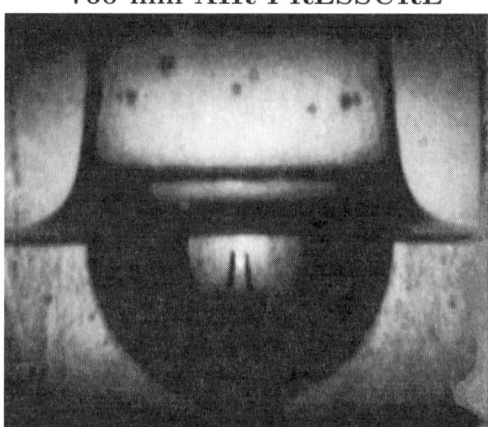

150 mm AIR PRESSURE

Fig. 6.13. The Kedrinskii results for an exploding bridge wire in water at 760 mm and 150 mm air pressure.

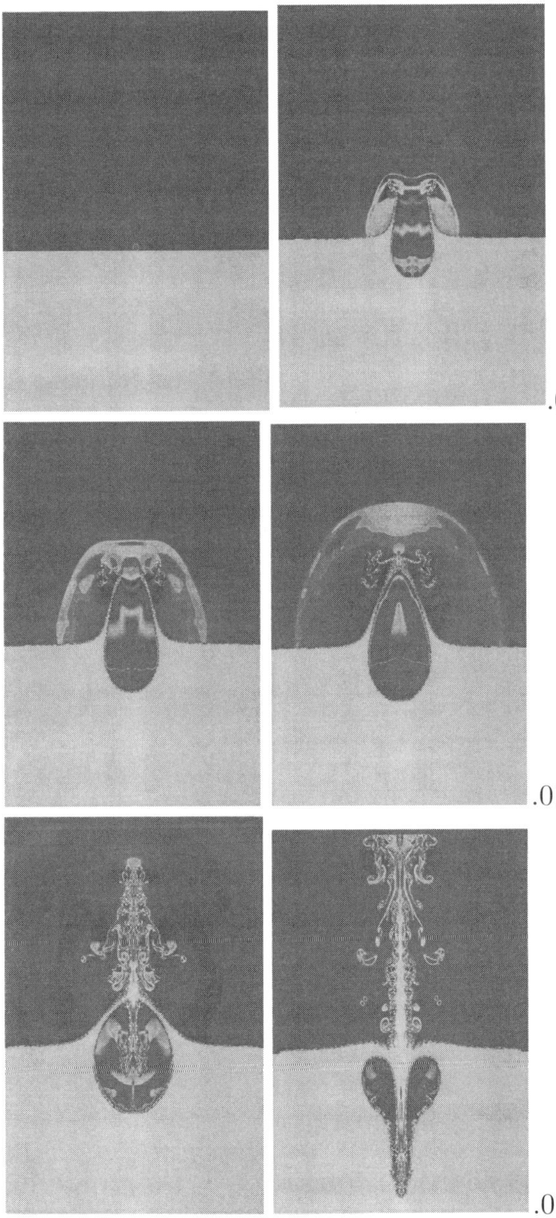

Sec. 6C WATER CAVITY GENERATION

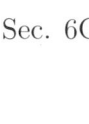

Fig. 6.14. The density profiles for a 0.25 cm diameter Aluminum projectile moving at 2.0 kilometers/sec impacting water at 5 atmosphere air pressure. The times are 0, 5, 10, 15, 30, and 70 milliseconds. The graphs are 100 cm wide by 120 cm tall, with 50 cm of water.

MODELING WAVES USING COMPRESSIBLE MODELS Chap. 6

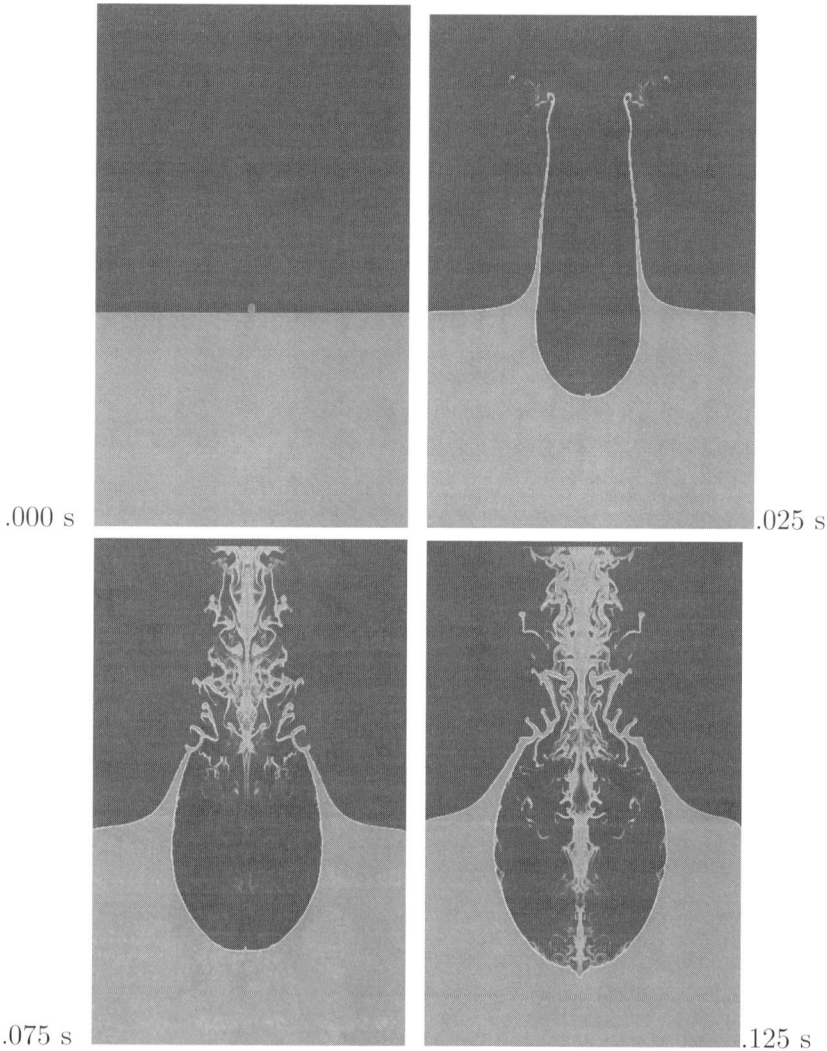

.000 s   .025 s
.075 s   .125 s

Fig. 6.15. The density profiles for a 0.25 cm diameter Aluminum projectile moving at 2.0 kilometers/sec impacting water at 1 atmosphere air pressure. The times are 0, 25, 75, and 125 milliseconds. The graphs are 100 cm wide and 120 cm tall, with 50 cm of water and 70 cm air.

Sec. 6C WATER CAVITY GENERATION

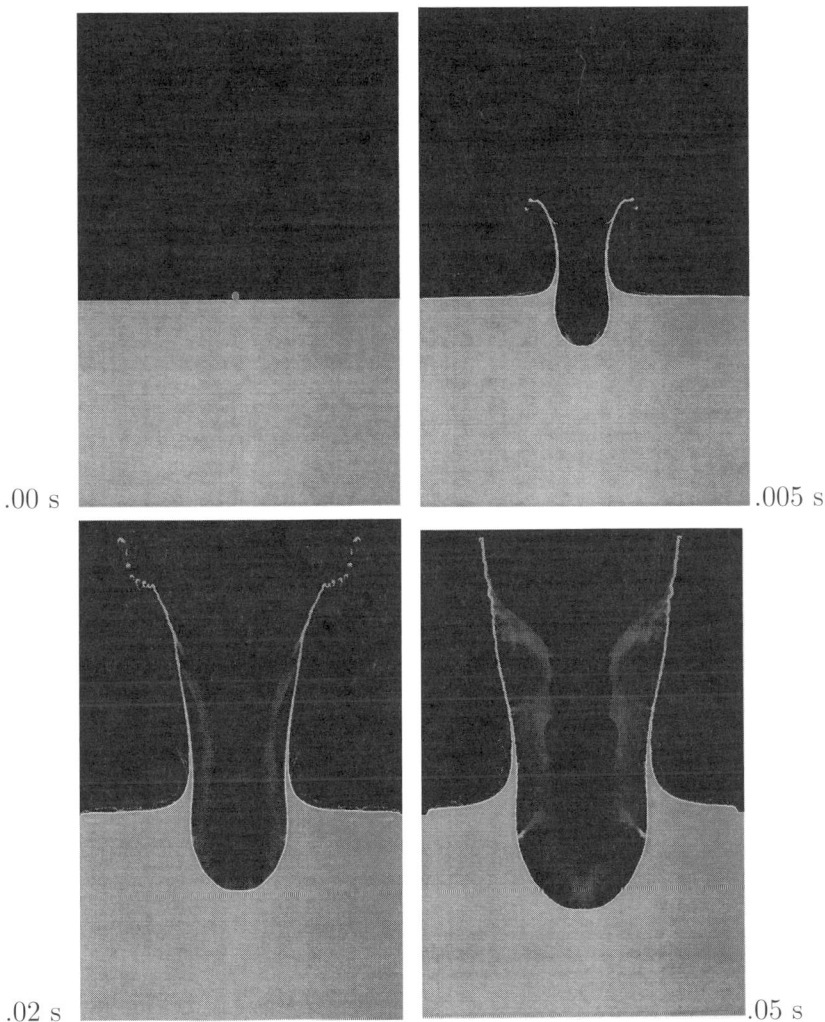

Fig. 6.16. The density profiles for a 0.25 cm diameter Aluminum projectile moving at 2.0 kilometers/sec impacting water at 76 mm (0.1 atmosphere) air pressure. The times are 0, 5, 20, and 50 milliseconds. The graphs are 80 cm wide by 100 cm tall with 40 cm of water and 60 cm air.

MODELING WAVES USING COMPRESSIBLE MODELS   Chap. 6

Explosive Generated Cavities

In reference 14, detailed one-dimensional compressible hydrodynamic modeling is described for explosives detonated in deep water. Agreement was obtained with the experimentally observed explosive cavity maximum radius and the period of the oscillation. It was concluded that the detonation product equation of state over the required range of 1 megabar to 0.01 atmosphere was adequate for accurately determining the energy partition between detonation products and the water. It was also concluded that the equations of state for water and detonation products were sufficiently accurate that they could be realiably used in multidimensional studies of water cavity formation and the resulting water wave generation in the region of the "upper critical depth".

The "upper critical depth" is the experimentally observed location of a explosive charge relative to the initial water surface that results in the maximum water wave height. It occurs when the explosive charge is approximately two-thirds submerged. The observed wave height at the upper critical depth is twice that observed for completely submerged explosive charges at the "lower critical depth." If the waves formed are shallow water waves capable of forming tsunamis, then the upper critical depth phenomenon would be important in evaluating the magnitude of a tsunami event from other than tectonic events.

The water wave amplitude as a function of the depth the explosive is immersed in water is shown in Figure 6.17. The scaled amplitude is AR/W and the scaled depth is D/W where A is maximum wave amplitude at a distance R from the explosive charge of weight W. The "upper critical depth" is at the first wave height maximum which occurs when an explosive charge is located at the water surface. The second smaller increase in wave height is at the "lower critical depth" which is about half the upper critical height but results in longer wave length water waves. Data are included for explosives with weights of 0.017 to 175 kilograms. The Craig[10] experimental results are shown with a large **x**.

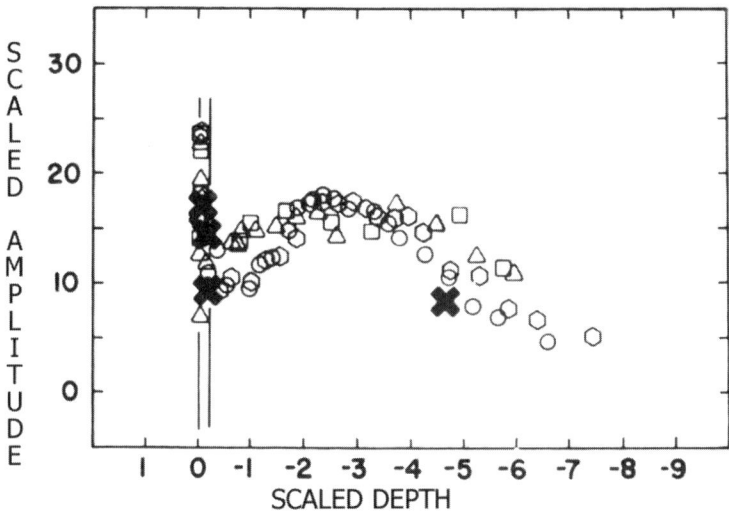

Fig. 6.17. The scaled wave height as a function of scaled explosive charge depth. The "upper critical depth" is the explosive charge depth when the maximum wave height occurs which is approximately two-thirds submerged. The second smaller increase in wave height is the "lower critical depth."

During the study of the upper critical depth phenomenon in the 1960's evidence of complicated and unexpected fluid flows during water cavity formation was generated by B. G. Craig and described in references 1, 10 and 14.

A sphere of explosive consisting of a 0.635 cm radius XTX 8003 (80/20 PETN/Silicon Binder at 1.5 g/cc) explosive and a 0.635 cm radius PBX-9404 (94/6 HMX/binder at 1.84 g/cc) explosive was detonated at its center. The sphere was submerged at various depths in water. PHERMEX[15] radiographs and photographs were taken with framing and movie cameras. The radiographs are shown in Figure 6.18.

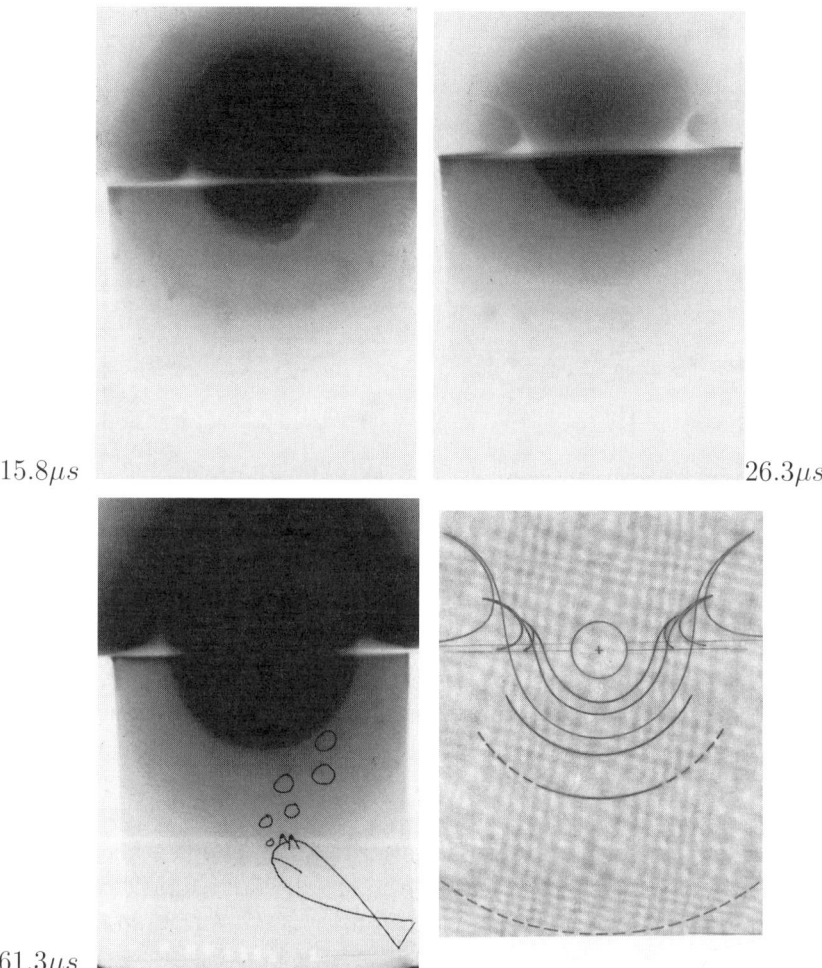

Fig. 6.18. Dynamic radiographs of a 2.54 cm diameter PBX-9404 explosive sphere detonated at its center and half submerged in water at 1 atmosphere air pressure. The times are 15.8, 26.3 and 61.3 microseconds. The sketch shows the prominent features of the radiographs with the water shock dashed.

The cavity, water ejecta and water surface profiles shown in the PHERMEX radiography in Figure 6.18 were closely approximated by the compressible hydrodynamic modeling described in reference 14 using the 2DE code and in Figure 6.19 using the *NOBEL* code.

Sec. 6C WATER CAVITY GENERATION

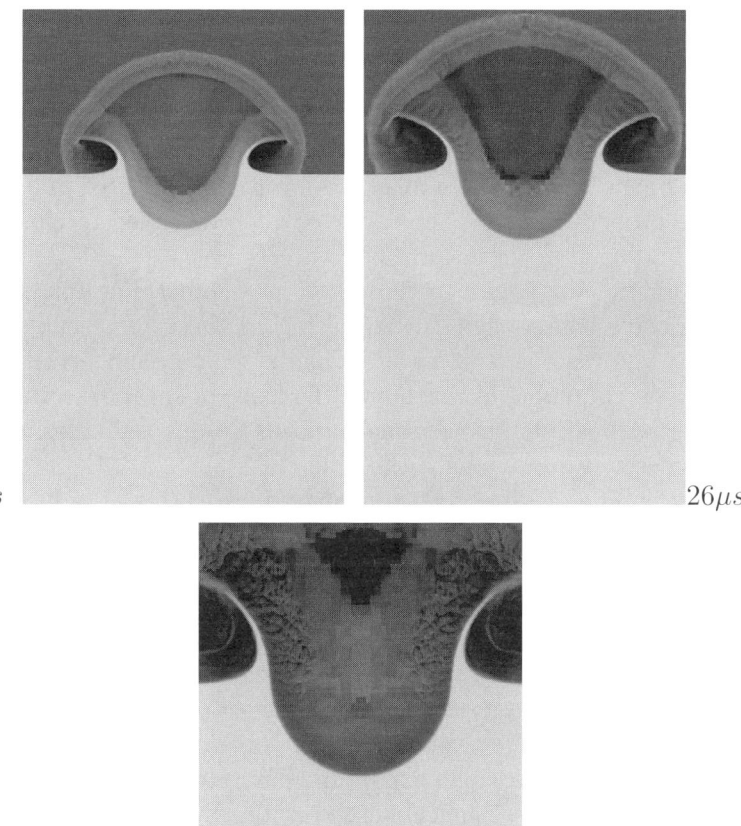

Fig. 6.19. *NOBEL* density profiles for a 2.54 cm diameter PBX-9404 explosive sphere detonated at its center and half submerged in water at 1 atmospheric air pressure. The times are 15.0, 26.0 and 60.0 microseconds for comparision with the radiographs in Figure 6.18. The width is 16 cm and the height is 24 cm, of which 16 cm is water.

MODELING WAVES USING COMPRESSIBLE MODELS

At later times, while the water cavity was expanding the upper ejecta plume collapsed and converged on the axis generating an upward water ejecta jet on the axis above the water plume and a downward water jet which generated a root on the axis below the bottom of the cavity. These results were not anticipated and neither was the observation that the water cavity proceeded to close at its top and descend down into deeper water.

At first it was assumed that there was something unique about the explosive source that was resulting in these remarkable observations. The reactive compressible hydrodynamic numerical models available were unable to reproduce the experimental observations or suggest any possible physical mechanisms unique to explosives.

As described previously, the different behavior of the water cavity as it expanded when the atmospheric pressure was reduced from one atmosphere to less than a third of an atmosphere is independent of the method used to generate the cavity, such as an exploding bridge wire or a hypervelocity projectile impact.

These remarkable experimental observations resisted all modeling attempts for over 25 years. The numerical simulations could not describe the thin water ejecta plumes formed above the cavity or the interaction with the atmosphere on the outside of the ejecta plume and the pressure inside the expanding cavity and plume.

Figure 6.20 shows the calculated water profiles for a 0.25 cm diameter PBX-9404 explosive sphere detonated at its center half submerged in water at one atmosphere air pressure.

All the projectile, exploding bridge wire and explosive experimental observations were reproduced as the atmospheric pressure was varied. At sufficiently high atmospheric pressure, the difference between the pressure outside the ejecta plume and the decreasing pressure inside the water plume and cavity as it expanded, resulted in the ejecta plume converging and colliding at the axis. A jet of water formed and proceeded above and back into the bubble cavity along the axis. The jet proceedes back through the bubble cavity penetrating the bottom of the cavity and forms the root observed experimentally. The complicated cavity collapse and resulting descent into deeper water was also numerically modeled.

Sec. 6C   WATER CAVITY GENERATION

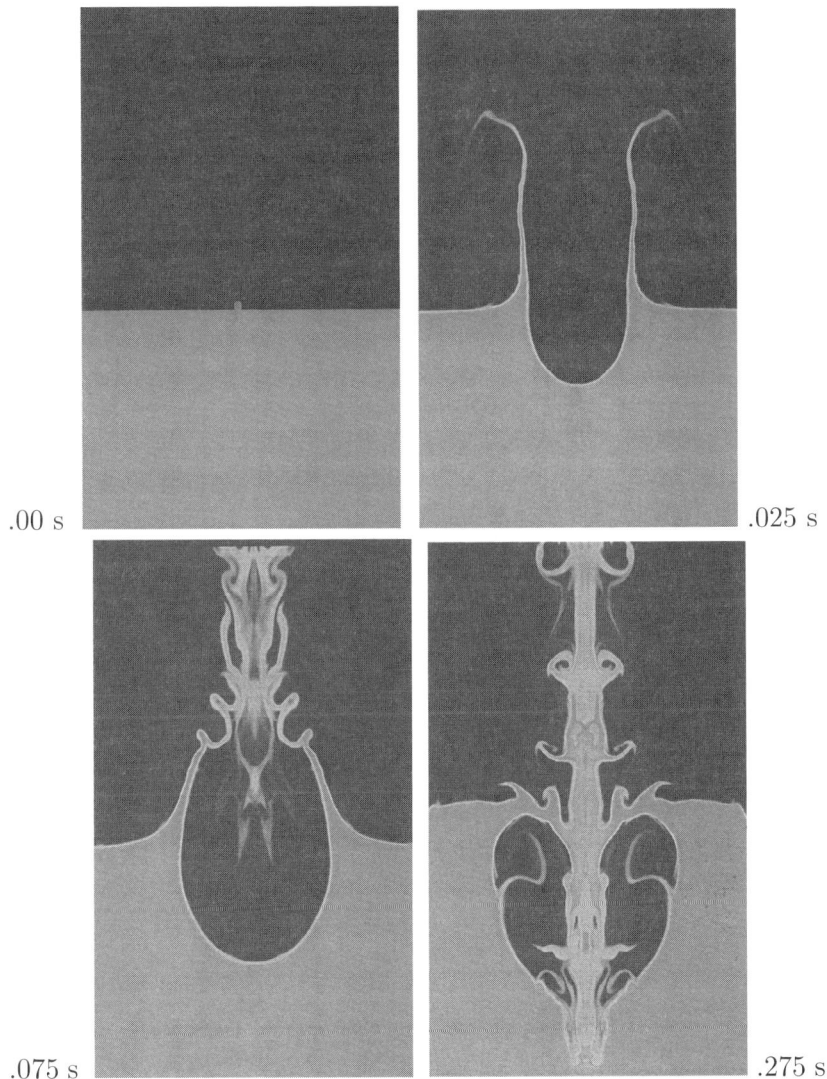

Fig. 6.20. The density profiles for a 2.54 cm diameter PBX-9404 explosive sphere detonated at its center and half submerged in water at 1 atmosphere air pressure. The times are 0, 25, 75, and 275 milliseconds. The graphs are 100 cm wide and 120 cm tall, with 50 cm of water and 70 cm of air.

MODELING WAVES USING COMPRESSIBLE MODELS

Chap. 6

Explosive Generated Water Wave

Craig[10] measured the wave amplitude as a function of time for the first few seconds at a distance of 4 meters from a 2.54 cm diameter PBX-9404 explosive sphere initiated at its center in 3 meters of water. He included mass markers in the water. Mass markers located 1 meter below the water surface and markers located 0.5 meter below the surface and 1 meter from the explosive showed no appreciable movement compared with those located nearer the surface or explosive charge. These results showed that the wave formed was not a shallow water wave.

The experimental and calculated wave parameters are summarized in Table 6.1. The parameters are given after 4 meters of travel from the center of the explosive charge. The wave parameters for the Airy wave were calculated using the *WAVE* code described in Chapter 1 for a depth of 3 meters and the experimentally observed wave length. Since the group velocity is almost exactly half the wave velocity, the Airy wave is a deep water wave. The shallow water results are from reference 14. A small wave from the initial cavity formation is followed by a larger negative and then a positive wave resulting from the cavity collapse. Only the second wave parameters are given in the table.

TABLE 6.1

Calculated and Experimental Wave Parameters

	Experimental	Airy Wave	Shallow Water	*NOBEL*
Wave Velocity (m/sec)	2.50±0.2	2.41	5.42	2.50±0.10
Amplitude (cm)	0.8	1.0	10.1	0.6
Wave Length (m)	3.75	3.75(input)	1.0	3.75
Period (sec)	1.5	1.55	0.18	1.5±0.1
Group Velocity (m/sec)		1.21		

Sec. 6C   WATER CAVITY GENERATION

The wave gauge was close to the edge of the water tank, which resulted in reflected waves which perturbed the subsequent wave measurements. Since the wave gauge was located close (0.69 meter) to the side of the tank, the reflections from the first small wave perturbed the second wave, which probably explains the larger than calculated amplitude, as the calculated wave was unperturbed by any boundary.

Conclusions

In the late 1960's and early 1970's, B. G. Craig at the Los Alamos National Laboratory reported observing the formation of ejecta jets and roots from cavities generated by small spherical explosives detonated near the water surface while the gas cavity was expanding.

The hypervelocity impact (1.25 to 6 kilometers/sec) of projectiles into water was studied at the University of Arizona in the early 1980's by Gault and Sonett. They observed quite different behavior of the water cavity as it expanded when the atmospheric pressure was reduced from one to a tenth atmosphere. Above about a third of an atmosphere, a jet of water formed above the expanding cavity and a root developed below the bottom of the bubble cavity. They did not occur for atmospheric pressures below a third of an atmosphere.

Similar results were observed in the middle 1980's by Kedrinskii at the Institute of Hydrodynamics in Novosibirsk, Russia when the water cavity was generated by exploding bridge wires, with jets and roots forming for normal atmospheric pressure and not for reduced pressures.

During the last decade, a compressible Eulerian hydrodynamic code called *SAGE* has been under development by the Los Alamos National Laboratory and Science Applications International (SAIC) which has continuous adaptive mesh refinement (AMR) for following shocks and contact discontinuities with a very fine grid.

A version of the *SAGE* code that models explosives, called *NOBEL*, has been used to model the experimental geometries of Sonett and of Craig. The experimental observations were reproduced as the atmospheric pressure was varied.

When the atmospheric pressure was increased, the difference between the pressure outside the ejecta plume above the water cavity and the decreasing pressure inside the water plume and cavity as it expanded resulted in the ejecta plume converging and colliding at the axis forming a jet of water proceeding above and back into the bubble cavity along the axis. The jet proceeding back through the bubble cavity penetrated the bottom of the cavity and formed the root observed experimentally. The complicated cavity collapse was numerically modeled.

Now that a code is available that can describe the experimentally observed features of projectile interaction with the ocean, we have a tool that can be used to evaluate impact landslide, projectile or asteroid interactions with the ocean, and the resulting generation of tsunami waves.

A PowerPoint presentation with Craig's and Sonett's experimental movies and computer animations is available on the NMWW CD-ROM in the NOBEL/CAVITY directory.

6D. Asteroid Generated Tsunamis

Introduction

Two- and three-dimensional compressible hydrodynamic modeling using the *SAGE* code has been performed to study the generation of tsunamis by asteroid impacts with the ocean[16,17].

The goal was to determine the characteristics of impact generated tsunami waves as a function of the size and energy of the asteroid. The evaluation in the popular press of the potential threats from modest sized earth crossing asteroids often overestimates the tsunami threat from small (100 to 250 meters in diameter) asteroids. The studies reported are based on faulty shallow water modeling, from which they conclude that the resulting short (less than 5 minutes) period tsunamis are a major threat throughout an ocean basin.

The news media are often delighted to promote such imaginary ocean basin wide "mega-tsunami" threats from small asteroid impacts, underwater landslides and volcanic or other explosions. Part of the problem is that until recently there has not been a method for realistically modeling such threats. That changed with the development of the *NOBEL/SAGE/RAGE* compressible AMR code described in this chapter.

While tsunamis up to a kilometer in initial height are generated by impactors of a kilometer diameter, they do not propagate as long period tsunamis such as generated by large earthquakes. Asteroid impact generated waves are highly complex in form and interact strongly with shocks propagating through the water and ocean crust. They decay more rapidly than the inverse of the distance from the impact point.

A panel of tsunami scientists of The Tsunami Society have evaluated the proposed tsunami hazards from events that generate initially high amplitude short period tsunamis. Their evaluation is available on The Tsunami Society web site at

http://www.sthjournal.org/media.htm.

MODELING WAVES USING COMPRESSIBLE   Chap. 6
MODELS

The Tsunami Society study concludes the following:

No mega-tsunami has occurred in either the Atlantic or Pacific oceans in recorded history. The colossal collapse of Krakatau or Santorin generated catastropic waves in the immediate areas but hazardous waves did not propagate to distance shores. Carefully performed numerical and experimental model experiments on such events verify that the relatively short period waves from these small, though intense, occurances do not travel as do tsunami waves from a major earthquake.

The 1958 Lituya Bay mega-tsunami described earlier in this chapter was the largest wave in recorded history, but it barely registered at nearby tide gauges. It had a short period and wavelength, which resulted in the wave rapidly decaying as it traveled out into the deep ocean.

Two-Dimensional Vertical Impact Asteroid Generated Tsunamis

A series of calculations in two-dimensions (cylindrical symmetry) for asteroids impacting the ocean vertically at 20 kilometers/sec was performed using the ASCI BlueMountain machine at the Los Alamos National Laboratory. These simulations were designed to follow the passage of an asteroid through the atmosphere, its impact with the ocean, the cavity generation, and subsequent re-collapse and generation of the tsunami. The results of these studies are described in reference 16.

The parameter study included different asteroid masses. Stony and iron bodies of diameters 250 meters, 500 meters and 1000 meters were used. The kinetic energies of the impacts ranged from 1.3 Gigatons to 195 Gigatons of equivalent TNT yield. The projectile moved through an exponential atmosphere into a 5 kilometers deep ocean. The impactors were composed of mockup mantle material (dunite with a density of 3.32 g/cc) or iron (density of 7.8 g/cc). as a mockup for nickel-iron asteroids. For these projectiles, analytical Mie-Gruneisen equations of state were used. An elastic-plastic strength model was used for the crust and asteroid materials. Water was described using the SAIC/Pactech tables. An example montage from the two-dimensional parameter study is shown in Figure 6.21.

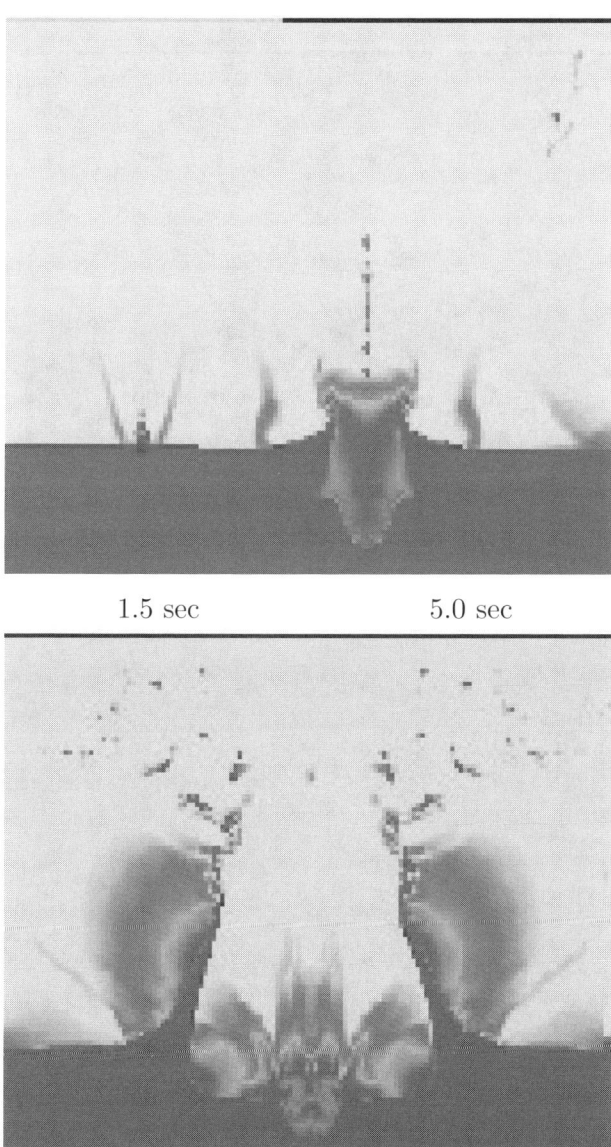

1.5 sec 5.0 sec

33.0 sec - Maximum Cavity

Fig. 6.21. A 1 kilometer iron vertical impactor craters the basalt crust, excavates a cavity in the ocean with a diameter of 25 kilometers.

MODELING WAVES USING COMPRESSIBLE Chap. 6
MODELS

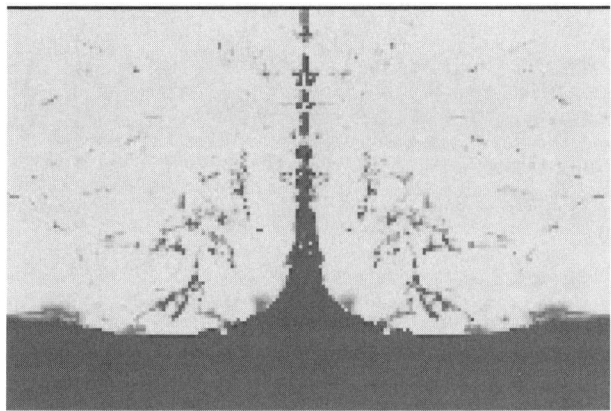

142.0 sec - Maximum Jet

Fig. 6.21 continued.

The density plot shortly after the collapse of the transient water crater is shown in Figure 6.22.

Fig. 6.22. The wave train density plots for a 1 kilometer iron asteroid impacting 5 kilometers of water and a basalt crust. The complex wave train is a result of the interaction of multiple shocks propagating through the water and the basalt crust.

Sec. 6D   ASTEROID GENERATED TSUNAMIS

The complex wave train is a result of the reflections and interaction of multiple shocks propagating through the water and basalt crust. A complicated time-dependent flow occurs which is best studied using the PowerPoint presentation and computer movies on the NMWW CD-ROM in the directory /NOBEL/ASTWAVE/.

It becomes obvious that our previous attempts to model the tsunami wave generation by asteroids using incompressible models were inadequate for the task. A tabular summary of the parameter study is presented in Table 6.2. Listed are the impact characteristics of the asteroid (composition, diameter, density, mass, velocity and kinetic energy) and the measured characteristics of the impact (maximum depth and diameter of the cavity, quantity of water displaced, time of maximum cavity, maximum jet and jet rebound, tsunami wave length and velocity).

TABLE 6.2

Asteroid Generated Tsunamis

Asteroid Material	Dunite	Iron	Dunite	Iron	Dunite	Iron
Asteroid Diameter	250 m	250 m	500 m	500 m	1000 m	1000 m
Asteroid Density	3.32 g/cc	7.81 g/cc	3.32 g/cc	7.81 g/cc	3.32 g/cc	7.81 g/cc
Asteroid Mass	2.7e13 g	6.4e13 g	2.2e14 g	5.1e14 g	1.7e15 g	4.1e15 g
Asteroid Velocity	20 km/s	20 km/s	20 km/s	20 km/s	20 km/s	20 km/s
Kinetic Energy	1.3 GT	3.1 GT	10.4 GT	24.4 GT	83 GT	195 GT
Max Cavity Dia	4.4 km	5.2 km	10.0 km	12.6 km	18.6 km	25.2 km
Max Cavity Depth	2.9 km	4.3 km	4.5 km	5.7 km	6.6 km	9.7 km
Water Displacement	4.4e16 g	9.1e16 g	3.5e17 g	7.1e17 g	1.8e18 g	4.8e18 g
Time of Max Cavity	13.5 s	16.0 s	22.5 s	28.0 s	28.5 s	33.0 s
Time of Max Jet	54.5 s	65.0 s	96.5 s	111 s	128 s	142 s
Time of Rebound	100.5 s	118.5 s	137.5 s	162 s	187 s	218 s
Tsunami Wavelength	9 km	12 km	17 km	20 km	23 km	27 km
Tsunami Velocity	120 m/s	140 m/s	150 m/s	160 m/s	170 m/s	175 m/s

The shallow water tsunami velocity in 5 kilometers deep water is 221 meters/sec.

The amount of water displaced during the formation of the cavity was found to scale linearly with the kinetic energy of the asteroid, as illustrated in Figure 6.23. A fraction of this displaced mass is actually vaporized during the explosive phase of the encounter, while the rest is pushed aside by the pressure of the vapor to form the crown and rim of the transient cavity.

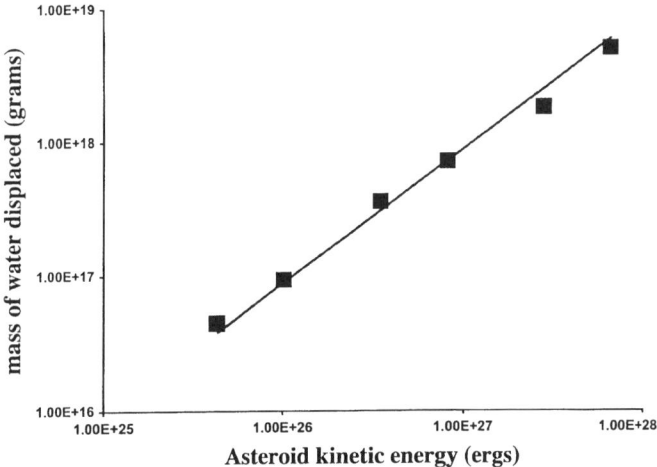

Fig. 6.23. The mass of water displaced in the initial cavity formation scales with the asteroid kinetic energy. The squares are the results from the parameter study tabulated in Table 6.2. The solid line shows direct proportionality. A fraction ($\approx$5–10 percent) of this mass is vaporized at impact.

The tsunami amplitude evolves in a complex manner, eventually decaying faster than the expected recipocal of the distance of propagation from the impact point as shown in Figure 6.24.

The wave trains are highly complex because of the multiple shock reflections and interactions involving the seafloor. The complexity of the wave breaking associated with the reflection and interaction of shocks results in a very dynamic maximum wave height. Realistic seafloor topography would undoubtedly influence the development of the water wave.

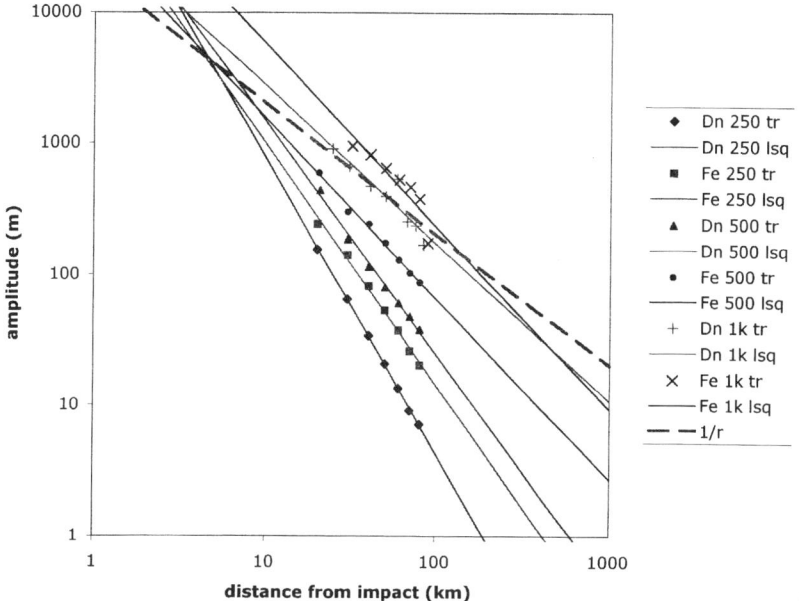

Fig. 6.24. The tsunami amplitude has a very complex early time profile. It eventually decreases with distance faster than 1/(Radius). The legend identifies the points associated with individual cases. The notation signifies the asteroid composition ("Dn" for dunite and "Fe" for iron) and the diameter in meters. A "tr" indicates calculated tracer heights and a "lsq" indicates a least square fitted line to the tracer heights.

For an ocean of 5 kilometers depth, the shallow water velocity is 221 meters/sec. In Figure 6.25 are shown the wave crest positions as a function of time for the simulations in the parameter study, along with constant velocity lines of 150 and 221 meters/sec. The wave velocities are substantially lower than the shallow water limit.

The tsunami wave length is found to scale roughly with the 1/4 power of the asteroid kinetic energy. The reason for this is that the wavelength is determined by the cavity-jet-rebound cycle. The time scale for this varies as $\sqrt{j^h/g}$ where j^h is the mean jet height and g is gravity. The mean jet height varies as the square root of the asteroid kinetic energy.

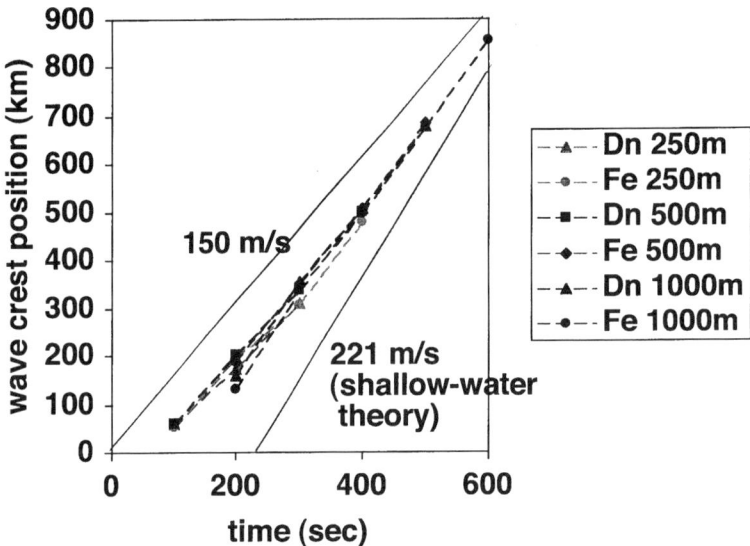

Fig. 6.25. The tsunami wave crest positions as a function of time for the parameter study cases. The notation is the same as for Figure 6.24. The solid lines of constant velocity illustrate that these waves are substantially slower than the shallow water prediction.

Three-Dimensional Oblique Impact Asteroid Generated Tsunamis

Three-dimensional simulations of a 1 kilometer diameter iron asteroid impacting the ocean at a 45 degree angle at 20 kilometers/sec were performed using the ASCI White machine. The calculations used up to 1200 processors and several weeks of computing time. Up to 200,000,000 computational cells were used, with a total computation time of 1,300,000 cpu-hours. The computational volume was a rectangular box 200 kilometers long in the direction of the asteroid trajectory, 100 kilometers wide and 60 kilometers tall. The height was divided into 42 kilometers of atmosphere, 5 kilometers of ocean water, 7 kilometers basalt crust and 6 kilometers of mantle material.

Sec. 6D   ASTEROID GENERATED TSUNAMIS

The asteroid was started at a point 30 kilometers above the surface of the water as shown in Figure 6.26. The atmosphere was a standard exponential atmosphere, so the air initially surrounding the asteroid was only 1.5 percent of the sea level density.

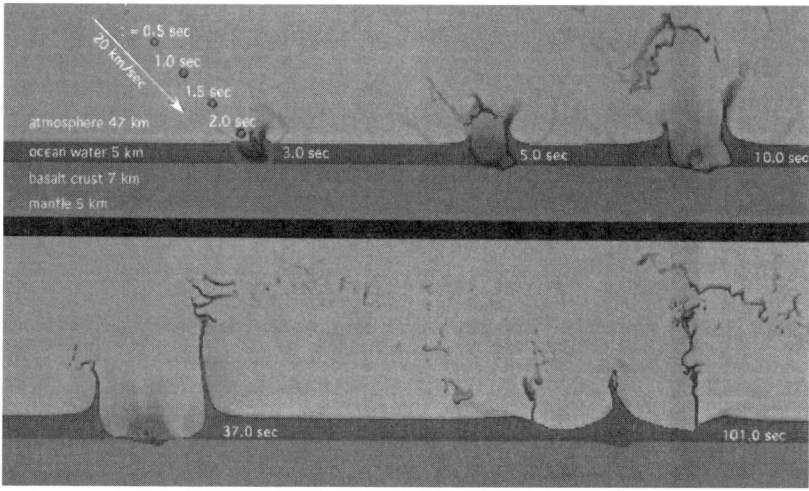

Fig. 6.26. A 1 kilometer diameter iron impactor at an angle of 45 degrees craters a 5 kilometers deep ocean and the basalt crust. Two-dimension slices in the vertical plane containing the asteroid trajectory are shown. The initial asymmetry disappears with time.

During the 2.1 sec of the asteroid's atmospheric passage at about Mach 60, a strong bow air shock develops, heating the air to temperatures of 1 electron volt (11,600 K). Less than 1 percent of the asteroid's kinetic energy is dissipated in the atmospheric passage.

The water is much more effective at slowing the asteroid, and essentially all of its kinetic energy is absorbed by the ocean and seafloor within 0.7 sec. The water immediately surrounding the trajectory is vaporized. The rapid expansion of the vapor cloud evacuates a cavity in the water that eventually expands to a diameter of 15 kilometers. This initial cavity is asymmetric because of the inclined trajectory of the asteroid. The splash, or crown, is markedly higher on the side opposite the incoming trajectory, as shown in Figure 6.27.

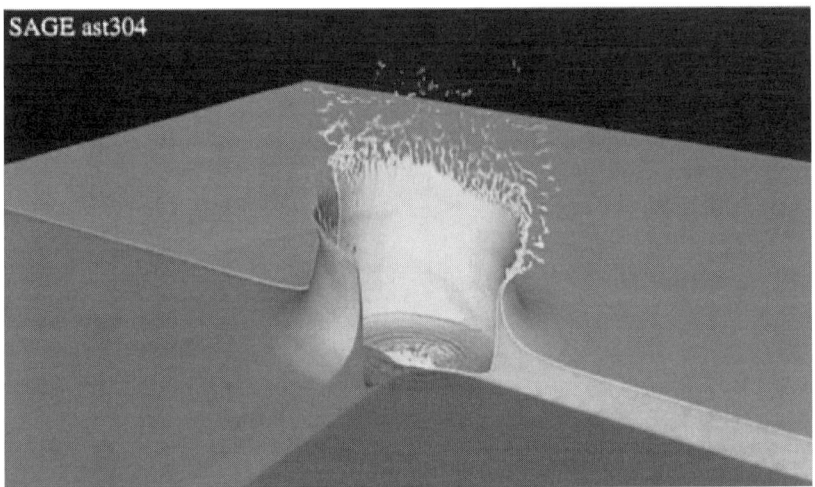

Fig. 6.27. A 1 kilometer diameter iron asteroid impacted the ocean from the right at 45 degrees. A perspective cutaway view from the same run as in Figure 6.26 at the time of maximum cavity (27.5 sec after impact) is shown. The density profiles are shown. The cut passes through the ocean to the basalt ocean floor. The diameter of the cavity is about 25 kilometers.

The maximum height of the crown on the downstream side is nearly 30 kilometers at 70 sec after impact. The collapse of the bulk of the crown makes a "rim wave" or precursor tsunami that propagates outward, somewhat higher on the downstream side. The higher portion of the crown breaks up into droplets that fall back into the water giving this precursor tsunami a very uneven and asymmetric profile.

The rapid dissipation of the asteroid's kinetic energy is very much like an explosion. Shocks propagate outward from the cavity in the water, in the basalt crust and in the mantle beneath. Multiple reflections of shocks and acoustic waves between the material interfaces complicate the dynamics.

The hot vapor from the initial cavity expands into the atmosphere, mainly downstream because of the momentum of the asteroid. When the pressure of the vapor in the cavity has diminished sufficiently, at about 35 sec after the impact, water begins to fill the cavity from the bottom.

Sec. 6D ASTEROID GENERATED TSUNAMIS

This filling has a high degree of symmetry because of the uniform gravity responsible for the water pressure. An asymmetric fill could result from non uniform seafloor topography, but that case was not modeled. The filling water converges on the center of the cavity and the implosion produces another series of shock waves, and a jet that rises vertically in the atmosphere to a height in excess of 20 kilometers at a time of 115 sec after impact, as shown in Figure 6.28. It is the collapse of this central vertical jet that produces the principal tsunami wave shown in Figure 6.29.

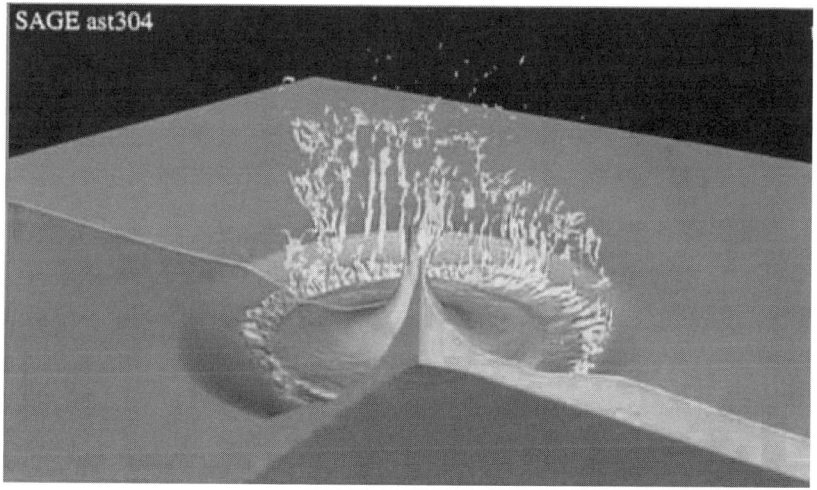

Fig. 6.28. Similar to Figure 6.27 but 115 sec after impact, which is the time of the formation of the central vertical jet. The collapse of the crown has produced a circular rim that is propagating outward, but the principal tsunami wave will be produced by the collapse of the central vertical jet. The crown splash has collapsed, pockmarking the surface of the first wave.

The evolution of this wave was modeled out to a time of 400 sec after impact. The inclined impact eventually produces a tsunami that is nearly circularly symmetric at late times, as shown in Figure 6.29. The tsunami has an initial height in excess of 1 kilometer, and decreases to 100 meters at a distance of 40 kilometers from the initial impact. Its propagation speed is 175 meters/sec, which is much less than the shallow water speed of 221 meters/sec.

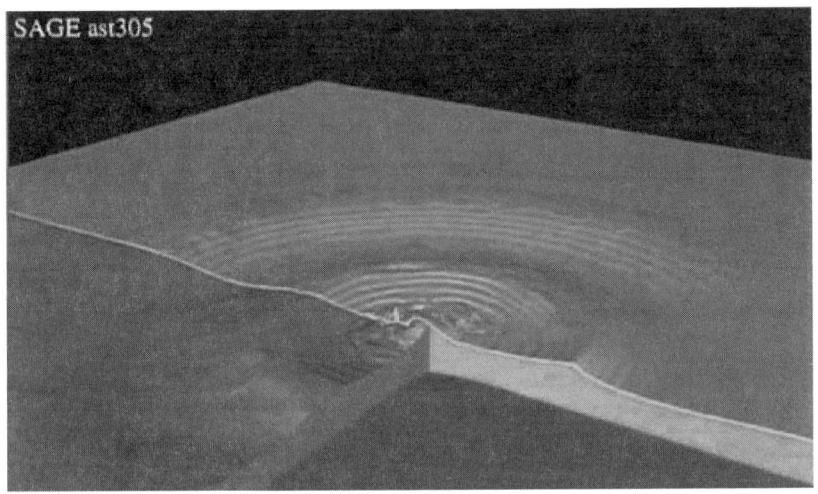

Fig. 6.29. Similar to Figure 6.27 and 6.28 but at 270 sec. The central jet has collapsed and formed the tsunami wave. The pock-marked precursor wave and the smoother principal wave are evident. The principal wave is about 1.5 kilometers in initial amplitude and moves with a speed of 175 meters/sec.

The evolution of this wave was modeled out to a time of 400 sec after impact. The inclined impact eventually produces a tsunami that is nearly circularly symmetric at late times as shown in Figure 6.29. The tsunami has an initial height in excess of 1 kilometer, and decreases to 100 meters at a distance of 40 kilometers from the initial impact. Its propagation speed is 175 meters/sec, which is much less than the shallow water speed of 221 meters/sec.

The 45 degree angle chosen for the three-dimensional simulation is the most probable angle for impacts. Impacts at 30 degrees and 60 degrees were also modeled. Significant differences with impact angle were found for the KT impactor described in reference 17 and in the next section of this chapter.

Conclusions

All asteroid impacts, including the oblique ones, produce a large underwater cavity with nearly vertical walls followed by a collapse starting from the bottom and subsequent vertical jetting. Substantial amounts of water are vaporized and lofted into the atmosphere. In the larger impacts, significant amounts of the crustal and even mantle material are lofted as well. Tsunamis up to a kilometer in initial height are generated by the collapse of the vertical jet. These waves are initially very complex in form and interact strongly with shocks propagating through the water and the crust. The tsunami waves were followed out to 100 kilometers from the point of impact. Their periods and wavelengths show them to be intermediate type waves and not shallow water waves. At great distances, the waves decay as the inverse of the distance from the impact point, ignoring sea floor topography. For all impactors smaller than about 2 kilometers diameter, the impacting body is highly fragmented and it remains lofted into the stratosphere with the water vapor and crustal material, hence very little trace of the impacting body will be found for most oceanic impacts. In the oblique impacts, the initial asymmetry of the transient crater and crown does not persist beyond a tsunami propagation length of 50 kilometers.

6E. KT Chicxulub Event

Introduction

The impact that created the Chicxulub crater in Mexico's Yucatan Peninsula is widely believed to be responsible for the mass extinctions at the end of the Cretceous period. This event has been extensively studied and modeled during the few years since its discovery. The stratigraphy at Chicxulub involves rocks like calcite and anhydrite that are highly volatile at pressures reached during impact. The volatility of the Chicxulub strata made the event very dangerous to the megafauna of the late Cretaceous.

On a geological time scale, impacts of asteroids and comets with the earth are a relatively frequent occurrence, causing significant disturbances to biological communities and strongly perturbing the course of evolution. Most famous among known catastrophic impacts is the one that marked the end of the Cretceous period and the dominance of the dinosaurs.

It is now widely accepted that the world sequence of mass extinctions at the Cretaceous-Tertiary (KT) boundary 65 million years ago was directly caused by the collision of an asteroid or comet with the earth. Evidence for this includes the large (200 kilometers diameter) buried impact structure at Chicxulub, Yucatan, Mexico, the world wide distributed Iridium layer at the KT boundary, and tsunami deposits well inland in North America, all dated to the same epoch as the extinction event.

As described in reference 17, the KT impactor was an asteroid of diameter roughly 10 kilometers. Its impact was oblique, either from the SE at 30 degrees to the horizontal or from the SW at 60 degrees. The asteroid encountered layers of water, anhydrite, gypsum, and calcium carbonate, which resulted in the lofting of many hundreds of cubic kilometers of these materials into the stratosphere, where they resided for many years and produced a global climate cooling that was fatal to many large animal species on earth.

Three-Dimensional KT Impact Model

The Chicxulub impact structure was discovered by scientists from Petroleos Mexicanos (Pemex), the Mexican national oil company[18,19]. It is believed to be the location of an impact that was responsible for the mass extinction at the end of the Cretaceous period, as proposed by Alvarez[20] et al. on the basis of the anomaly in abundances of Iridium and other Platinum group elements in the KT boundary bedding plane.

Paleogeographic data suggest that the crater site, which presently straddles the Yucatan coastline, was submerged at the end of the Cretaceous period on the continental shelf. The substrate consisted of fossilized coral reef over continental crust.

Sec. 6E KT CHICXULUB EVENT

In the Gisler, et al.[17] simulation, the multilayered target was 300 meters of water, 3 kilometers of calcite, 30 kilometers of granite and 18 kilometers of mantle material. Above the target was an exponential atmosphere up to 106 kilometers altitude. The asteroid's plunge was numerically started at the 106 kilometers altitude. Three-dimensional simulations were performed with impact angles of 30, 45 and 60 degrees to the horizontal. The computational domain extended 256 kilometers by 128 kilometers and simulated the half space of 256 by 256 kilometers.

The calculations were performed on the new ASCI Q computer at Los Alamos, a cluster of ES45 Alpha boxes from HP/Compaq. The problems used 1024 processors and a total of about 1 million cpu hours. The adaptive mesh included up to a third of a billion computation cells.

Three prominent features of the simulation are illustrated by the 45 degree impact shown in Figure 6.30.

Fig. 6.30. The density at seven sec after a 10 kilometers diameter granite asteroid impacts the earth at 45 degrees. The isosurface, at density 0.005 g/cc is chosen to show everything denser than air. The scale is set by the back boundary which is 256 kilometers long. The maximum height of the uplifted material is 50 kilometers.

After impact, billions of tons of very hot material are lofted into the atmosphere. A water plume directed away from the impact angle carries much of the horizontal component of the asteroid's momentum in the downrange direction. This material, consisting of vaporized fragments of the projectile mixed with the target, is extremely hot, and will ignite vegetation many hundreds of kilometers away from the impact site.

The highly turbulent and energetic ejecta plume is directed predominatly upward as shown in Figure 6.31.

Fig. 6.31. The density at 42 sec after a 10 kilometers diameter granite asteroid impacts the earth at 45 degrees. The isosurface, at density 0.005 g/cc is chosen to show everything denser than air. The scale is set by the back boundary which is 256 kilometers long. The maximum height of the uplifted material is 50 kilometers.

Some material is projected into orbits that terminate far outside the computational volume. The dissipation of the kinetic energy, some 300 teratons TNT equivalent, produces a stupendous explosion that melts, vaporizes and ejects a substantial volume of water and granite.

Ballistic trajectories carry some of this material back to earth in a conical debris curtain that gradually moves away from the crater lip and deposits a blanket of ejecta around the forming crater as shown in Figure 6.32. The debris curtain has separated from the rim of the still forming crater as material in the curtain falls to earth. The debris from the curtain is deposited in a blanket of ejecta that is asymmetric around the crater with more in the downrange than in the uprange direction. The distribution of material in the ejecta blanket was used as a diagnostic to determine the direction and angle of impact of the asteroid. Some material is projected into orbits that terminate far outside the computational volume.

Fig. 6.32. The density at two minutes after a 10 kilometers diameter granite asteroid impacts the earth at 45 degrees. The isosurface, at density 0.005 g/cc is chosen to show everything denser than air. The scale is set by the back boundary which is 256 kilometers long. The debris curtain has separated from the crater.

The blanket of ejecta is found to be strongly asymmetrical around the crater, with the uprange portion much thinner than the rest. This is a result of the coupling of the horizontal component of the asteroid's momentum to the debris, and to the ionized and shocked atmosphere in the asteroid's wake.

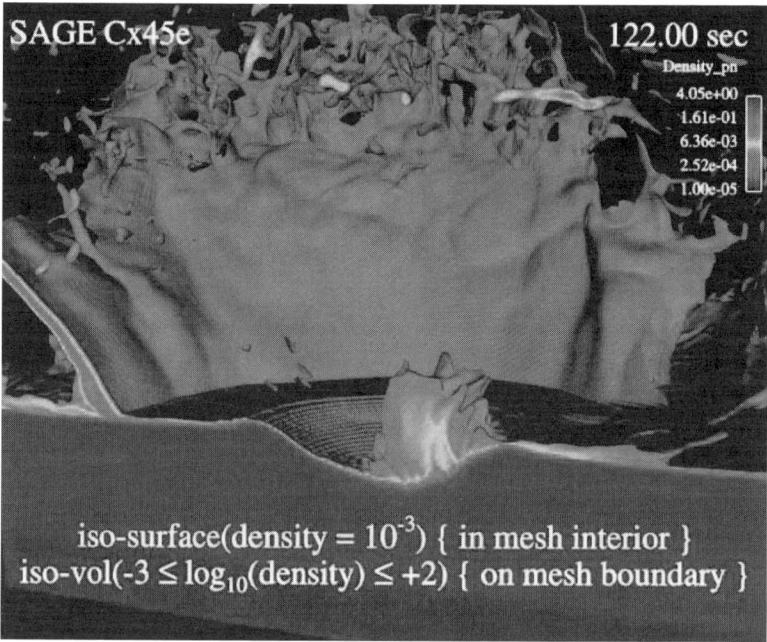

Fig. 6.33. The density at 122 sec after a 10 kilometers diameter granite asteroid impacts the earth at 45 degrees. A cut through the center of the impact cavity shows that no water is present from the center of the crater out to the debris curtain.

As shown in Figure 6.33, the ocean water is mostly vaporized in the crater region and beyond the crater to the debris curtain which has separated from the rim of the crustal crater. The initial shallow 300 meters deep water where the asteroid impact occurred would not support the generation of major long period tsunami waves. However, the ocean water is driven by strong shock waves and the expanding curtain out into surrounding deep ocean. This would result in large, long period tsunamis that would propagate around the world and leave the tsunami deposits well inland in North America, as observed during the KT epoch.

Spectacular computer movies of the KT impact and a PowerPoint presentation produced by Dr. Galen Gisler are on the NMWW CD-ROM in the directory /NOBEL/KTIMPACT/.

Sec. 6E KT CHICXULUB EVENT

Dr. Gisler performed the KT asteroid impact calculations as part of the Los Alamos Crestone Project validation studies. The *NOBEL/SAGE/RAGE* hydrodynamic codes and the 3-D graphics were developed by the Crestone Project team. The Chicxulub PowerPoint presentation has been shown to many members of the legislative and executive branches of the U. S. government and to Los Alamos visiting dignitaries including Prince William.

If life on earth is to continue we must develop the capability of deflecting asteroids before they collide with the earth. Solem[21] has described the options available for deflection of asteroids and comets. He concludes "the most cost effective and the ONLY currently available means of asteroid disruption (deflection or pulverization) is a nuclear explosive." So the Los Alamos nuclear weapon program that has developed most of the water wave modeling technologies described in this book as part of its nuclear weapon effects studies also has developed the nuclear technology that is both a major threat to life on earth and that which makes it possible to defend the earth from the extinction of life by asteroids and comets.

References

1. Charles L. Mader, *Numerical Modeling of Explosives and Propellants,* CRC Press, Boca Raton, Florida (1998).
2. M. L. Gittings, "1992 SAIC's Adaptive Grid Eulerian Code," Defense Nuclear Agency Numerical Methods Symposium, pp. 28–30 (1992).
3. R. L. Holmes, G. Dimonte, B. Fryxell, M. L. Gittings, J. W. Grove, M. Schneider, D. H. Sharp, A. L. Velikovich, R. P. Weaver and Q. Zhang, "Richtmyer-Meshkov Instability Growth: Experiment, Simulation and Theory," Journal of Fluid Mechanics, Vol. 9, pp. 55–79 (1999).
4. R. M. Baltrusaitis, M. L. Gittings, R. P. Weaver, R. F. Benjamin and J. M. Budzinski, "Simulation of Shock-Generated Instabilities," Physics of Fluids, Vol. 8, pp. 2471–2483 (1996).
5. Don J. Miller, "Giant Waves in Lituya Bay, Alaska" Geological Survey Professional Paper 354-C, U. S. Government Printing Office, Washington, D.C. (1960).
6. Frances E. Calwell, *Land of The Ocean Mists–The Wild Ocean Coast West of Glacier Bay*, Alaska Northwest Publishing Company, Edmonds, Washington (1986).
7. Charles L. Mader, "Modeling the 1958 Lituya Bay Tsunami," Science of Tsunami Hazards, Vol. 17, pp. 57–67 (1999).
8. Hermann M. Fritz, Willi H. Hager and Hans-Erwin Minor, "Lituya Bay Case: Rockslide Impact and Wave Runup," Science of Tsunami Hazards, Vol. 19, pp. 3–22 (2001).
9. Charles L. Mader, "Modeling the 1958 Lituya Bay Mega-Tsunami, II ," Science of Tsunami Hazards, Vol. 20, pp. 241–250 (2002).
10. Bobby G. Craig, "Experimental Observations of Underwater Detonations Near the Water Surface," Los Alamos Scientific Laboratory report LA-5548-MS (1972).
11. Donald E. Gault and Charles P. Sonett, "Laboratory Simulation of Pelagic Asteroid Impact: Atmospheric Injection, Benthic Topography, and the Surface Wave Radiation Field," Geological Society of America, Special Paper 190, pp. 69–92 (1982).

REFERENCES

12. Valery Kedrinskii, private communication (1985).
13. Charles L. Mader and Michael L. Gittings, "Dynamics of Water Cavity Generation," Science of Tsunami Hazards, Vol. 21, pp. 91–118 (2003).
14. Charles L. Mader, *Numerical Modeling of Detonations,* University of California Press, Berkeley, California (1979).
15. Charles L. Mader, Timothy R. Neal and Richard D. Dick, *LASL Phermex Data, Volume I, II, and III* , University of California Press, Berkeley, California (1980). Available at http://lib-www.lanl.gov/ladcdmp/ph1.pdf.
16. Galen Gisler, Robert Weaver, Michael L. Gittings and Charles Mader, "Two- and Three-Dimensional Simulations of Asteroid Ocean Impacts," Science of Tsunami Hazards, Vol. 21, pp. 119–134 (2003).
17. Galen Gisler, Robert Weaver, Michael L. Gittings and Charles Mader, "Two- and Three-Dimensional Asteroid Impact Simulations," Computers in Science and Engineering (2004).
18. A. R. Hildebrand, G. T. Penfield, D. A. Kring, M. Pilkington, Z. A. Camargo, S. B. Jacobsen and W. V. Boynton, "Chicxulub Crater: A Possible Cretaceous/Tertiary Boundary Impact Crater on the Yucatan Peninsula, Mexico." Geology, Vol. 19, pp. 867–871 (1991).
19. V. L. Sharpton, G. B. Dalrymple, L. E. Marin, G. Ryder, B. C. Schuraytz and J. Urruita-Fucugauchi, "New Links Between the Chicxulub Impact Structure and the Cretaceous/Tertiary Boundary," Nature, Vol. 359, pp. 819–821 (1992).
20. L. W. Alvarez, W. Alvarez, F. Asaro and H. V. Michel, "Extraterrestrial Cause for the Cretaceous/Teritary Extinction," Science, Vol. 208, pp. 1095–1108 (1980).
21. Johndale C. Solem, "Comet and Asteroid Hazards: Threat and Mitigation ," Science of Tsunami Hazards, Vol. 17, pp. 141–154 (1999).

CD-ROM CONTENTS

Tsunami Animations

A collection of tsunami animations performed using the *SWAN* code described in the *Numerical Modeling of Water Waves - Second Edition,* by Dr. Charles L. Mader. A few calculations were performed using the full Navier-Stokes *ZUNI* or *SOLA* codes. The tsunami animations are archived at

http://t14web.lanl.gov/Staff/clm/tsunami.mve/tsunami.htm.

To view animation type or click – MOVIE – To study individual cases type – MM TAPEXX where XX is file number. Follow instructions on screen.

1960.MVE - May 23, 1960 tsunami generation in Chile, propagation across the Pacific Ocean, and indundation of Hilo, Hawaii. Described in "Modeling Hilo, Hawaii Tsunami Inundation," Science of Tsunami Hazards, Vol. 9, pp. 85–94 (1991), and Scientific Computing and Automation, June issue, pp. 19–23 (1993).

1964.MVE - Tsunami of April 1, 1964 generation in Gulf of Alaska, propagation across the Pacific Ocean, and inundation of Crescent City, California. See "Tsunami Inundation Model Study of Eureka and Crescent City, California," NOAA Tech. Memo. ERL PMEL-105 (1994).

60THEAT.MVE - The interaction of the tsunami of May 23, 1960 with the Hilo, Hawaii Theater. Described in PACON 1993.

90HILO.MVE - 1990 Hilo topography and buildings inundated by a 1960 tsunami wave. See also HOTEL.MVE.

2ATAST.MVE - The inundation of the U.S. East Coast by a 100 meters, 2000 sec tsunami wave that could be generated by an asteroid.

10NYAST.MVE - The inundation of the U.S. East Coast by a wave from the incompressible collapse of a 10 kilometers radius cavity, 3000 meters deep and a 100 kilometers radius cavity in the Atlantic Ocean off New York City.

AIMPACT.MVE - An impact cavity collapse and tsunami generation study using shallow water and full Navier-Stokes models. Described in "Modeling Asteroid Impact and Tsunami," Science of Tsunami Hazards, Vol. 16, pp. 21–30 (1998).

CD-ROM CONTENTS

ATLAST.MVE - A tsunami in the Atlantic Ocean generated by the incompressible collapse of a cavity 150 kilometers wide and 3500 meters deep.

AUSAST.MVE - Interaction of a tsunami with Australia from a Hawaii landslide generated tsunami and from a cavity collapse generated tsunami. Described in "Modeling of Tsunami Propagation Directed at Wave Erosion on Southeastern Australia Coast 105,000 Years Ago," Science of Tsunami Hazards, Vol. 13, pp. 45–52 (1995).

BBAY.MVE - A study of the vulnerability of Berau Bay, Indonesia to tsunamis.

CASCAD.MVE - Inundation of U.S. West Coast by a tsunami from the Cascadia fault.

ECAST.MVE - The inundation of the U.S. East Coast by a tsunami generated by the incompressible collapse of a 150 kilometers wide, 3000 meters deep cavity. See also NYAST.MVE

ELTAST.MVE - Described in "Modeling the Eltanin Asteroid Tsunami," Science of Tsunami Hazards, Vol. 16, pp. 17–20 (1998).

EURAST.MVE - The inundation of Europe by a 100 meters high and 2000 sec period tsunami.

EUREKA.MVE - The Eureka, California tsunami of April 25, 1992. See "Tsunami Inundation Model Study of Eureka and Crescent City, California," NOAA Tech. Memo. ERL PMEL-105 (1994).

GUS.MVE - The Furumoto sources for the Hawaiian tsunamis of 1946, 1957, 1964 and 1965. Part of a source modeling project for Dr. A. Furumoto, Hawaii Civil Defense Tsunami Advisor.

HIAST.MVE - The inundation of the Hawaiian Islands by a 100 meters high, 2000 sec period tsunami wave. Described in "Asteroid Tsunami Inundation of Hawaii," Science of Tsunami Hazards, Vol. 14, pp. 85–88 (1996).

HILAND.MVE - The tsunami generated by a landslide off the Kona coast of the island of Hawaii about 105 Ka years ago. Described in "Modeling the 105 Ka Landslide Lanai Tsunami," Science of Tsunami Hazards, Vol. 12, pp. 33–38 (1994).

HKAI.MVE - Inundation of Hawaii Kai, Hawaii by a typical off shore 3 meters high, 1500 sec tsunami wave.

HOTEL.MVE - The interaction of a May 23, 1960 tsunami wave with current Hilo, Hawaii tourist hotels. See also 90HILO.MVE.

CD-ROM CONTENTS

HUMBOL.MVE - Tsunami inundation of Humboldt Bay, California by an offshore maximum expectable 10 meters high, 2000 sec tsunami wave. See "Tsunami Inundation Model Study of Eureka and Crescent City, California," NOAA Tech. Memo. ERL PMEL-105 (1994).

ICEAST.MVE - The inundation of Iceland by a 100 meters high and 2000 sec period tsunami.

INDIA.MVE - Tsunami in the Indian Ocean generated by the incompressible collapse of a cavity 38 kilometers wide and 4000 meters deep.

INDONES.MVE - Indonesia tsunami of December 12, 1992.

JAPAST.MVE - The inundation of Tokyo, Japan by a tsunami generated by a incompressible cavity collapse. Described in "Asteroid Tsunami Inundation of Japan," Science of Tsunami Hazards, Vol. 16, pp.11–16 (1998).

KAIAKA.MVE - Tsunami inundation of Kaiaka Bay, Oahu, Hawaii by the 1952 tsunami.

KBAY.MVE - Tsunami inundation of Kaneohe Bay, Hawaii by a typical offshore 3 meters high, 2000 sec tsunami and by a maximum expectable offshore 10 meters high, 2000 sec tsunami wave.

KONA.MVE - Tsunami inundation of Kona, Hawaii by a typical offshore 3 meters high, 2000 sec tsunami wave.

KURIL.MVE - The tsunami of October 1994 generated off the Kuril islands of Japan.

LAAST.MVE - Inundation of Los Angeles, California by a 100 meters high, 2000 sec period tsunami wave.

LAPALMA.MVE - Modeling the proposed La Palma landslide tsunami. Published in "Modeling the La Palma Landslide Tsunami," Science of Tsunami Hazards, Vol. 19, pp. 160–180 (2001).

LAUP.MVE - The April 1, 1946 tsunami inundation of Laupahoehoe, Hawaii.

LITUYA.MVE - The July 8, 1958 mega-tsunami at Lituya Bay, Alaska with inundations up to 520 meters. Described in "Modeling the 1958 Lituya Bay Mega-Tsunami," Science of Tsunami Hazards, Vol. 17, pp. 57–67 (1999). The Lituya Bay impact landslide generation of the tsunami is described in Chapter 6 and in Science of Tsunami Hazards, Vol. 20, pp. 241–250 (2002).

CD-ROM CONTENTS

LISBON.MVE - Modeling the 1755 Lisbon tsunami generation and propagation across the Atlantic Ocean to the Caribbean. Science of Tsunami Hazards, Vol. 19, pp. 93–98 (2001).

LOIHI.MVE - A study using the *ZUNI* full Navier-Stokes code of the tsunami wave generation and propagation from the collapse of the Loihi, Hawaii summit in August, 1996.

M9CALIF.MVE - An M9 earthquake generated tsunami interacting with the Oregon and California coasts.

NIC.MVE - The tsunami generated off the coast of Nicaragua in 1992. Described in "Modeling the 1992 Nicaragua Tsunami," Science of Tsunami Hazards, Vol. 11, pp. 107–110 (1993).

NYAST.MVE - The inundation of the U.S. East Coast by the incompressible collapse of a 100 kilometers radius 3000 meters deep cavity. Another tsunami wave had a height of 100 meters and a 2000 sec period. See also 10NYAST.MVE.

ORAST.MVE - A 100 meters high, 2000 sec period tsunami interacting with the Oregon coast.

OREGM9.MVE - An M9 earthquake generated tsunami interacting with the Oregon coast.

PACAST.MVE - Tsunami in the middle of the Pacific formed from the incompressible collapse of a cavity 150 kilometers wide and 4500 meters deep.

PROP.MVE - Described in "Numerical Tsunami Propagation Study," Science of Tsunami Hazards, Vol. 11, pp. 93–106(1993) and in Chapter 5 of *Numerical Modeling of Water Waves - Second Edition*.

SANDY.MVE - Tsunami inundation of Sandy Beach region of Oahu, Hawaii by a typical offshore 3 meters high, 2000 sec tsunami and by a maximum expectable offshore 10 meters high, 2000 sec tsunami wave.

SANFAST.MVE - Inundation of San Francisco, California by a 100 meters high, 2000 sec tsunami wave.

SKAGWAY.MVE - The landslide generated tsunami of November 3, 1994 at Skagway, Alaska. The Skagway modeling is described in "Modeling the 1994 Skagway Tsunami," Science of Tsunami Hazards, Vol. 15, pp. 41–48 (1997). See also SOLA.MVE.

CD-ROM CONTENTS

SMSFAST.MVE - Inundation of San Francisco by a tsunami wave generated by the incompressible collapse of a 20 kilometers wide, 3000 meters deep cavity.

SOLA.MVE - Three-dimensional, full Navier-Stokes modeling using the MCC *SOLA* code of the November 3, 1994 Skagway, Alaska tsunami. See also SKAGWAY.MVE.

SOURCE.MVE - Described in "Numerical Tsunami Source Study," Science of Tsunami Hazards, Vol. 11, pp.81–92 (1993) and in Chapter 5 of *Numerical Modeling of Water Waves - Second Edition.*

VSLIDE.MVE - A landslide generated tsunami from the Chain of Craters road region of the island of Hawaii.

WAIANAE.MVE - The inundation of the leeward side of Oahu, Hawaii by a maximum expectable offshore 10 meters high, 2000 sec tsunami wave.

WAIPIO.MVE - The interaction of the May 23, 1960 tsunami with the Waipio, Hawaii region. The 50 foot inundation is the largest recorded in Hawaii.

WALKER.MVE - An evaluation of the vulnerability of Hawaii to tsunamis generated south of Honolulu, either along the Kona Coast or in the Tonga trench. Modeling requested by Dr. D. Walker, Oahu Civil Defence Tsunami Advisor.

WINDWARD.MVE - Tsunami inundation of the Windward side of Oahu, Hawaii by a typical offshore 3 meters high, 2000 sec tsunami and by a maximum expectable offshore 10 meters high, 2000 sec tsunami wave.

CD-ROM CONTENTS

NOBEL **PowerPoint Presentations**

A collection of PowerPoint presentations describing water wave studies performed using the compressible hydrodynamic code *NOBEL*. The studies are described in Chapter 6 of *Numerical Modeling of Water Waves - Second Edition* and archived at http://t14web.lanl.gov/Staff/clm/tsunami.mve/tsunami.htm.

The PowerPoint presentations may be viewed using PPVIEW in /NOBEL/PPRESENT/.

LITUYA - The July 8, 1958 Lituya Bay, Alaska impact landslide tsunami generation. A mega-tsunami was generated that reached an altitude of 520 meters. Laboratory experiments and numerical modeling results are presented. Described in "Modeling the 1958 Lituya Bay Mega-Tsunami, II," Science of Tsunami Hazards, Vol. 20, pp. 241–250 (2002).

CAVITY - The generation of cavities in water by projectile impacts and by explosives is described both experimentally and using compressible hydrodynamic models. Described in "Dynamics of Water Cavity Generation," Science of Tsunami Hazards, Vol. 21, pp. 91–118 (2003).

ASTWAVE - The generation of tsunamis by the impact of a 0.25 to 1 kilometer diameter asteroid at 20 kilometers/sec with 5 kilometers of ocean and 5 kilometers of basalt is modeled using compressible hydrodynamics in two and three dimensions. Described in "Two- and Three-Dimensional Simulations of Asteroid Ocean Impacts," Science of Tsunami Hazards, Vol. 21, pp. 119–134 (2003).

KTIMPACT - The KT Chicxulub asteroid impact event is modeled using the three-dimensional compressible Navier-Stokes model. Described in "Two- and Three-Dimensional Asteroid Impact Simulations," Computers in Science and Engineering (2004).

CD-ROM CONTENTS

CD-ROM CODE DIRECTORIES

WAVE – The *WAVE* code described in Chapter 1 solves the equations for Airy, third-order Stokes and Laitone solitary gravity waves. The directory contains the FORTRAN source code, the executable code for DOS or Windows and WAVE.PDF which describes the code.

SWAN – The shallow water *SWAN* code described in Chapter 2 solves the long wave, shallow water, nonlinear equations of fluid flow. The directory contains the FORTRAN source and executable codes which generate a graphics file that may be processed using the programs included. It also includes a description of the input to the code in the file SWAN.PDF. Examples and topographic files are furnished.

ZUNI – The incompressible Navier-Stokes *ZUNI* code described in Chapter 3 solves the incompressible, viscous fluid flows with a free surface using the Navier-Stokes equations. A detailed description of the computer program and its input file is included in the file ZUNI.PDF. The FORTRAN source and the executable codes are included.

SOLA – The incompressible three-dimensional Navier-Stokes *ZUNI* code described in Chapter 4 solves the incompressible viscous fluid flows with a free surface using the Navier-Stokes equations. The FORTRAN source and the executable codes are included. The Skagway 1994 tsunami is used as an example.

LGW – The Carrier linear gravity wave *LGW* code described in Chapter 5 uses analytical methods for solving the linear gravity model. The FORTRAN source and executable codes are included. Examples of Gaussian tsunamis described in Chapter 5 are furnished.

TIDE – A classic computer program for calculating tides with the FORTRAN source and executable codes furnished.

CD-ROM CONTENTS

Science of Tsunami Hazards Directory

All the *Science of Tsunami Hazards* journals through 2003 in PDF format may be searched using Adobe Acrobat 4.0 or higher. Issues of the journal are archived at http://epubs.lanl.gov/tsunami and current issues are available at http://www.sthjournal.org.

    Dir = TS211.PDF, TS212.PDF, TS213.PDF, TS214.PDF
    * Volume 21 (No. 1), (No. 2), (No. 3), (No. 4) 2003
    Dir = TS201.PDF, TS202.PDF, TS203.PDF, TS204.PDF, TS205.PDF
    * Volume 20 (No. 1), (No. 2), (No. 3), (No. 4), (No. 5), 2002
    DIR = TS191.PDF, TS192.PDF, TS193.PDF
    * Volume 19 (No. 1) (No. 2) (No. 3) , 2001
    DIR = TS181.PDF, TS182.PDF
    * Volume 18 (No. 1) (No. 2) , 2000
    DIR = TS171.PDF, TS172.PDF, TS173.PDF
    * Volume 17 (No. 1) (No. 2) (No. 3), 1999
    DIR = 00394718.PDF
    * Volume 16 (No. 1), 1998
    DIR = 00394719.PDF, 00394720.PDF
    * Volume 15 (No. 1) (No. 2), 1997
    DIR = 00394721.PDF, 00394722.PDF, 00394723.PDF
    * Volume 14 (No. 3) (No. 2) (No. 1), 1996
    DIR = 00394724.PDF
    * Volume 13 (No. 1), 1995
    DIR = 00394725.PDF, 00394726.PDF
    * Volume 12 (No. 2) (No. 1), 1994
    DIRECTORY = 00394727.PDF, 00394728.PDF
    * Volume 11 (No. 2) (No. 1), 1993
    DIRECTORY = 00394729.PDF
    * Volume 10 (No. 1), 1992
    DIRECTORY = 00394730.PDF, 00394731.PDF
    * Volume 9 (No. 2) (No. 1), 1991
    DIRECTORY = 00394732.PDF, 00394733.PDF
    * Volume 8 (No. 2) (No. 1), 1990

CD-ROM CONTENTS

DIRECTORY = 00394734.PDF, 00394735.PDF
* Volume 7 (No. 2) (No. 1), 1989
DIRECTORY = 00394736.PDF
* Volume 6 (No. 1), 1988
DIRECTORY = 00394737.PDF, 00394738.PDF
* Volume 5 (No. 2) (No. 1), 1987
DIRECTORY = 00394739.PDF, 00394740.PDF, 00394741.PDF
* Volume 4 (No. 3) (No. 2) (No. 1), 1986
DIRECTORY = 00394742.PDF
* Volume 3 (No. 1), 1985
DIRECTORY = 00394743.PDF, 00394744.PDF
* Volume 2 (No. 2) (No. 1), 1984
DIRECTORY = 00394745.PDF
* Volume 1 (No. 1), 1982

CNMWW.PDF is a searchable PDF file of the book *Numerical Modeling of Water Waves - Second Edition* with many figures in color.

AUTHOR INDEX

Numbers in *italic* type indicate pages where references are listed in full.

Alvarez, L. W. , 246, *253*
Alvarez, W. , 246, *253*
Amsden, A. A. , 100, *133*
Asaro, F. , 246, *253*

Baltrusaitis, R. M. , 202, *252*
Batchelor, G. K. , 1, *30*
Benjamin, R. F. , 202, *252*
Bernard, E. , 70, *95*
Bornhold, B. D. , 78, *95*
Boynton, W. V. , 246, *253*
Bretschneider, C. L. , 186, *197*
Budzinski, J. M. , 202, *252*
Butler, H. L. , 52, *94*

Calwell, F. E. , 205, *252*
Camargo, Z. A. , 246, *253*
Campbell, B. , 78, 83, 84, 90, *95*
Carrier, G. F. , 165, 169, *197*
Chan, R. K. C. , 100, 107–109, 111, *133*
Chubarov, L. B. , 37, 43, *92*
Cox, D. C. , 154, *164*
Craig, B. G. , 126, 130, *134*, 215, 218, 219, 225, 230–232, *252*
Curtis, G. D. , 37, 45, 62, 70, *93*, *94*, *95*

Dalrymple, G. B. , 246, *253*
Daly, B. J. , 99, 109, *133*
Dick, R. D. , 225, *253*
Dimonte, G. , 202, *252*
Divoky, D. , 52, 57, 59, *94*
Dronkers, J. J. , 33, *92*

AUTHOR INDEX

Fritz, H. M. , 208, 212, *252*
Fromm, J. E. , 108, 109, 111, *133*
Fryxell, B. , 202, *252*
Fuchs, R. A. , 121, 123, *134*
Fucugauchi, J. U. , 246, *253*
Furumoto, A. S. , 47, *93*

Garcia, A. W. , 52, *94*
Garcia, W. J. , 153, *163*
Gault, D. E. , 215, 216, 231, *252*
Gisler, G. , 233, 234, 244, 246, 247, 250, 251, *253*
Gittings, M. L. , 130, *134*, 201, 202, 215, 233, 234, 244, 246, 247, *252, 253*
Green, C. K. , 47, *93*
Greenspan, H. P. , 169, *197*
Grove, J. W. , 202, *252*

Hadi, S. , 42, *93*
Hager, W. H. , 208, *252*
Hansen, W. , 31, 34, *92*
Harlow, F. H. , 99, 100, 109, *133*
Hildebrand, A. R. , 246, *253*
Hirt, C. W. , 100, 107, 109, *133*, 135, *163*
Holmes, R. L. , 202, *252*
Houston, J. R. , 52, *94*
Hwang, L. , 52, 57, 59, *94*

Jacobsen, S. B. , 246, *253*
Johnson, J. W. , 121, 123, *134*

Kedrinskii, V. , 215, 217, 220, 231, *253*
Kinsman, B. , 1, *30*
Kowalik, Z. , 37, 43, 60, 88, 89, *92, 94, 95*
Kring, D. A. , 246, *253*
Kulikov, E. A. , 78, *95*

AUTHOR INDEX

Landau, L. D. , 1, *30*
Lander, J. F. , 77, 78, *95*
Le Mehaute, B. , 9, *30*
Lee, J. J. , 78, 79, 89, 90, *95*
Leenderstse, J. J. , *92*
Lifshitz, E. M. , 1, *30*
Loomis, H. G. , 153, 154, *164*
Lukas, S. , 43, *93*

Mader, C. L. , *30*, 37, 42, 43, 45, 62, 70, *93*,
  94, *95*, 103, 108, 125, *133*, *134*, 153,
  164, 165, *197*, 199, 201, 207, 211, 215,
  224–226, 230, 233, 234, 244, 246, 247,
  252, *253*
Madsen, O. S. , 113, 115, *133*
Marchuk, A. G. , 37, 43, *92*
Marin, L. E. , 246, *253*
Mei, C. C. , 113, 115, *133*
Michel, H. V. , 246, *253*
Miller, D. J. , 204, 207, 208, *252*
Minor, H. E. , 208, *252*
Moore, D. W. , 165, *197*
Morison, J. R. , 121, 123, *134*
Murty, T. S. , 37, 43, *92*

Nabeshima, G. , 62, *94*
Neal, T. R. , 225, *253*
Nichols, B. D. , 100, 107, 109, 135, 153, *163*, *164*
Nottingham, D. , 78, 83, 84, 90, 91, *95*

Pelinovsky, E. N. , 37, 43, *93*
Penfield, G. T. , 246, *253*
Peregrine, D. H. , 113, *133*
Petroff, C. , 78, 79, 89, 90, *95*
Pilkington, M. , 246, *253*
Plafker, G. , 52, 57, 72, *94*
Proudman, J. , 33, *92*

AUTHOR INDEX

Rabinovich, A. B. , 78, *95*
Raichlen, F. , 79, 89, 90, *95*
Ramming, H. G. , 37, 43, *92*
Romero, N. C. , 135, *163*
Ryder, G. , 246, *253*

Satake, K. , 70, *95*
Savage, J. C. , 57, *94*
Schneider, M. , 202, *252*
Schuraytz, C. , 246, *253*
Shannon, J. P. , 99, 100, 109, *133*
Sharp, D. H. , 202, *252*
Sharpton, V. L. , 246, *253*
Shokin, I. I. , 37, 43, *92*
Sklarz, M. A. , 153, *164*
Sokolowski, T. J. , 60, *94*
Solem, J. C. , 251, *253*
Sonett, C. P. , 215, 218, 219, 231, 232, *252*
Spielvogel, L. Q. , 153, *164*
Stein, L. R. , 135, *163*
Street, R. L. , 100, 107–109, 111, *133*

Tangora, R. E. , 153, *164*
Thomson, R. E. , 78, *95*

Uusitalo, S. , 34, *92*

Van Dorn, W. G. , 52, 53, 72, *94*
Velikovich, A. L. , 202, *252*
Vitousek, M. , 43, *93*

Watts, P. , 78, 79, 89, 90, *95*
Weaver, R. P. , 202, 233, 234, 246, 247, *252*, *253*
Welander, P. , 34, *92*
Welch, J. E. , 99, 109, *133*
Whalin, R. W. , 52, *94*
Whitmore, P. M. , 60, *94*
Wiegel, R. L. , 20, 22, *30*
Wybro, P. G. , 186, *197*

Zhang, Q. , 202, *252*

SUBJECT INDEX

1946, April 1 Tsunami, 45–59, 71–75, 256, 257
1958, July 8, Lituya Bay Tsunami, 203–214
1960 Tsunami Type, 65–67, 256
1960, May 23 Tsunami, 45, 46, 55–62, 65, 66, 94, 173, 181, 255, 259
1964, March 28 Tsunami, 45, 46, 52–59, 70–76, 92–94, 173, 181, 255, 256
1975, November 29, Hawaii Tsunami, 153–161
1994, November 3, Skagway Tsunami, 77–91, 163

Accuracy, 13, 37, 125, 141–143, 152
Airy Wave, 2, 3, 16–19, 29, 42, 110, 165, 167, 169, 177, 178, 183
AMR, 211, 214, 217, 231, 233
ASCI, 199, 234, 240, 247
Asteroid, 207, 215, 216, 232–234, 236–239, 245–260

Barrier, 120–124
Beach, 93, 100, 133, 197, 258
Berau Bay, 256
BlueMountain, 234
Bore, 9, 12, 45, 55, 59, 62
Bottom Motion, 36, 38
Boundary Layer, 9
Bridgewire, 216, 228
Brigham Victory Ship, 52
Bubble Cavity, 228, 231, 232
Bubble Radius, 125
Building, 45, 46, 56, 60–66

Cartesian, 31, 97, 136, 137, 141
Chicxulub, 245, 246, 251, 253, 260
Comet, 246, 251
Compressible Flow, 1, 3, 125, 175
Compressible Model, 197, 200–202
Conservation of Energy, 1, 6
Conservation of Momentum, 1, 5, 100, 141
Continental Slope, 42, 43, 108, 110–118
Continuative Boundary, 149, 150

SUBJECT INDEX

Continuum Boundary, 42
Convergence Criterion, 109, 110, 126
Coriolis, 28, 32, 33, 38, 41, 44, 45, 46
Crestone Project, 251

DeChezy, 35, 38, 49, 57, 189
Deep Water Wave, 129, 131, 132, 162, 230
Detonation, 125, 130, 215, 224
Dockslide, 81, 83, 88
Donor Cell, 141–143, 152

Earth Rotation, 32, 33
Earthquake, 45, 47, 52, 55, 57, 59, 60, 70–72, 75, 94,
  153–155, 161, 171, 173, 181, 197, 203,
  204, 206, 207, 233
Estuaries, 9, 42, 93
Eulerian, 3, 5, 6, 12, 16, 100, 199, 201, 211, 214,
  217, 231, 252
Explosion, 9, 125, 129, 130, 197, 199, 215, 233, 242, 248
Explosive Generated, 217, 224, 230
Extinction of Life, 251

Flood Wave, 9, 12
Flooding, 1, 4, 7, 9, 36, 37, 42, 44–76, 175, 182–196
FORTRAN, 261
Free Slip, 100, 148

Gravity, 2, 31, 33, 109, 126, 159, 202, 207, 211, 239, 243
Gravity Waves, 9, 29, 108, 119

Hawaiian Islands, 45, 47, 53, 57, 59, 256
Hilo Bay, 45, 46, 49, 52, 53, 57, 59, 66, 68
Hilo Harbor, 46, 51, 52, 53, 57, 59, 93
Hilo Pier No. 1, 52
Hilo Theater, 60–66, 255
Hotel, 45, 66, 68, 255, 256
Hydrostatic, 17, 33, 36, 110, 126

SUBJECT INDEX

Impact Landslide, 201, 212, 214, 232, 257, 260
Impact Landslide Experiment, 208–211
Incompressible Flow, 99, 133, 148, 199, 215
Indonesia, 42, 256
Intermediate Wave, 13, 16, 245

Jet, 130, 131, 215–217, 228, 231, 232, 236, 239, 243, 244, 245
JIMAR, 93, 164

Kinetic Energy, 129, 130, 237–239, 241, 242, 248
KT Impact, 244, 246, 250, 251, 260

Laitone Solitary Waves, 19, 20, 21, 29, 261
Landslide, 43, 78, 81–95, 154–164, 197, 199, 201, 207–214, 232, 233, 256–260
Linear Gravity Waves, 131, 165, 169–181, 261
Lituya Bay, 201–209, 214, 234, 252, 257, 260
Loihi, 130–132
Long Wave Paradox, 36, 37
Los Alamos, 1, 247, 251, 252
Los Alamos National Laboratory, 133, 163, 215, 231, 234, 254

MAC, 99, 100, 104, 107–109, 133, 135, 141, 151
Marker and Cell, 99, 135, 153
Mega-tsunami, 203, 214, 233, 234, 257, 260
Momentum, 103, 139, 141, 146, 148–150, 152, 248, 249
Musi Upang, 42, 93

National Academy of Scicnce, 94
Navier-Stokes, 1, 42, 44, 60, 71, 90, 97, 99, 101, 108, 125, 126 128, 129, 131, 132, 135, 137, 153, 154, 158–161, 165, 172, 173, 175, 178–183, 191, 194, 195–197, 211, 217, 258–261
News Media, 233
NMWW CD-ROM, 17, 20, 22, 29, 37, 38, 45, 65, 66, 76, 88, 91 130, 131, 163, 169, 182, 212, 232, 237, 250
No-slip, 148, 151
NOAA, 46, 52, 70, 71, 79, 83, 95, 255, 256, 259

SUBJECT INDEX

NOBEL, 125, 201, 211, 212, 214, 215, 217, 226, 227, 230, 231–233, 237, 250, 251, 260, 261
Nomenclature, 2, 38, 200
NTHMP, 91
Nuclear, 197, 251

Oahu, 257–259
Obstacle, 12, 100, 147, 151
Oscillatory Wave, 9

PARN Dock, 77, 78, 82, 85, 87, 90
PBX-9404 Explosive, 125, 225–230
PEMEX, 246
PHERMEX, 225, 226, 253
Piston Boundary, 42, 64, 65, 113, 115, 118
Potential Energy, 13
Potential Function, 10, 101, 105
PowerPoint Presentation, 130, 212, 232, 237, 251, 260
Progressive Waves, 11
Projectile, 51, 215–217, 231, 234

Q Computer, 247

Radiograph, 225–227
RAGE, 201, 233, 251
Reflective Boundary, 42, 154
Reynolds Number, 104
Rigid Wall, 103–106, 148
Root, 130, 215–218, 228, 231, 232, 239
Runup, 47, 100, 109, 111, 115, 119, 183, 185–196, 252

SAGE, 201, 202, 214, 231, 233, 255
SAIC, 231, 234, 252, 254
Scotch Cap Lighthouse, 47, 48
Sea Floor Displacement, 43
Sea of Japan, 43
Seiche, 9, 12

SUBJECT INDEX

Shallow Water, 12–18, 22, 31, 33, 37, 43–60, 71, 75, 80, 90, 91, 93, 100, 107–111, 115–130, 154–178, 180–182, 185, 187, 188, 190, 194–196, 201, 207, 215, 224, 230, 233, 239, 240, 243–245, 255
Shoaling, 90, 91, 132, 135, 137, 139, 142, 143, 149, 151, 154, 156, 163, 255, 258, 259, 261
Shock, 202, 216, 226, 238, 241, 243, 250, 252
Skagway, 77–80, 84, 85, 88, 90, 91, 95
SMAC, 102–105, 107, 109, 110, 113
SOLA-3D, 90, 91, 132, 135, 137, 139, 142, 143, 149, 151, 154, 156, 163, 255, 258, 259, 261
Source, 37, 44–59, 72, 75, 85, 94, 130, 153–164, 171–175, 181, 193, 196, 197, 228, 259, 261
Stability, 109, 151, 152
Standing Wave, 11, 12
Stationary Wave, 11
Steady State, 10, 14, 15, 16
Stem, 216, 217
Stokes Wave, 22, 23, 29, 261
Storm Wave, 9
Submerged Barrier, 120
Surface Marker Particles, 100, 107
SWAN, 33, 37, 42, 43, 45, 46, 59, 71, 75, 80, 88, 91, 93, 115, 125–128, 154, 156, 163, 165–167, 172, 175–177, 183, 185–188, 190, 193–195, 207, 255, 261

Three-dimensional, 1, 3, 26, 90, 91, 132, 135, 153, 154, 160–163, 182, 196, 200, 233, 240, 246, 253, 259, 260
Tide Gauge, 47, 52, 77, 79, 80, 82, 87–90, 95, 132
Tide Generating Force, 32, 33
Time Step, 38, 41, 61, 81, 109, 126, 131, 139, 145, 151, 152, 183, 195, 202, 207, 209
Translatory Waves, 9
Transmission Coefficient, 121, 123, 124
Tsunami Museum, 56
Two-dimensional, 28, 35, 42, 97, 108, 131, 153, 169, 175, 202, 234

SUBJECT INDEX

Underwater Barrier, 123
Unimak Island, 47, 48
Upper Critical Depth, 125, 129, 139, 224, 225

Viscosity, 2, 4, 6, 7, 34, 109, 126, 137, 152, 159, 167, 177, 189
Volcanic, 130, 197, 233
Vorticity, 101, 105, 109

Waianae Harbor, 43, 93, 259
Water Cavity, 2, 23, 128, 134, 201, 202, 215–217, 224, 228, 231, 253, 255, 260
WAVE Code, 17, 20, 22, 25, 29, 93, 230, 261
Wave Energy, 43, 120
Wave Length, 2, 11, 13, 14, 20, 22, 29, 33, 111, 114, 115, 118, 120, 130, 131, 163, 173, 175, 181, 186, 196, 224, 230, 237, 239
Wave Number, 2, 11
Wave Period, 2, 11, 47, 52, 88, 111, 121, 131, 173, 181, 183, 186–188, 193, 195
White Machine, 240
Wind Waves, 35

XTX-8003 Explosive, 225

ZUNI, 99, 100, 108–110, 115, 125, 126, 128, 129, 131–133, 165, 167, 168, 172, 175–178, 183, 189–195, 255, 258, 261

LIMITED WARRANTY

CRC Press LLC warrants the physical disk(s) enclosed herein to be free of defects in materials and workmanship for a period of thirty days from the date of purchase. If within the warranty period CRC Press LLC receives written notification of defects in materials or workmanship, and such notification is determined by CRC Press LLC to be correct, CRC Press LLC will replace the defective disk(s).

The entire and exclusive liability and remedy for breach of this Limited Warranty shall be limited to replacement of defective disk(s) and shall not include or extend to any claim for or right to cover any other damages, including but not limited to, loss of profit, data, or use of the software, or special, incidental, or consequential damages or other similar claims, even if CRC Press LLC has been specifically advised of the possibility of such damages. In no event will the liability of CRC Press LLC for any damages to you or any other person ever exceed the lower suggested list price or actual price paid for the software, regardless of any form of the claim.

CRC Press LLC specifically disclaims all other warranties, express or implied, including but not limited to, any implied warranty of merchantability or fitness for a particular purpose. Specifically, CRC Press LLC makes no representation or warranty that the software is fit for any particular purpose and any implied warranty of merchantability is limited to the thirty-day duration of the Limited Warranty covering the physical disk(s) only (and not the software) and is otherwise expressly and specifically disclaimed.

Since some states do not allow the exclusion of incidental or consequential damages, or the limitation on how long an implied warranty lasts, some of the above may not apply to you.

DISCLAIMER OF WARRANTY AND LIMITS OF LIABILITY

The author(s) of this book have used their best efforts in preparing this material. These efforts include the development, research, and testing of the theories and programs to determine their effectiveness. Neither the author(s) nor the publisher make warranties of any kind, express or implied, with regard to these programs or the documentation contained in this book, including without limitation warranties of merchantability or fitness for a particular purpose. No liability is accepted in any event for any damages, including incidental or consequential damages, lost profits, costs of lost data or program material, or otherwise in connection with or arising out of the furnishing, performance, or use of the programs in this book.